British
inner
cities

Hugh Matthews

OXFORD UNIVERSITY PRESS

ACKNOWLEDGEMENTS

The publishers and author would like to thank the following people for their permission to use copyright material:

p.4 *top* Maggie Murray, Format Picture Agency, *middle* Stuart Franklin, Magnum Photos, *bottom* Chris Steele-Perkins, Magnum Photos; p.9 Martine Franck, Magnum Photos; p.16 City of Nottingham, Planning Department; p.27 London Docklands Development Corporation; p.30 J. Allan Cash Agency; p.32 Jasphot Agency; p.33 Magnum Photos; p.34 London Docklands Development Corporation; p.35 London Docklands Development Corporation; p.36 John Mowlem and Co. plc; p.38 *left* Maggie Murray, Format Picture Agency, *right* J. Allan Cash Agency.

The cover photograph is reproduced by permission of Martine Franck, Magnum Photos.

Illustrations are by Herb Bowes Graphics, Oxford.

Every effort has been made to trace and contact copyright holders, but this has not always been possible. We apologise for any infringement of copyright.

For my parents

Oxford University Press, Walton Street, Oxford OX2 6DP

Oxford New York Toronto Delhi Bombay Calcutta Madras Karachi Petaling Jaya Singapore Hong Kong Tokyo Nairobi Dar es Salaam Cape Town Melbourne Auckland and associated companies in Berlin Ibadan

Oxford is a trade mark of Oxford University Press

© Oxford University Press 1991
First published 1991
Reprinted 1992

ISBN 0 19 913367 0

Typeset by Gem Publishing Company, Wallingford
Design and artwork by Herb Bowes Graphics, Oxford
Printed in Great Britain by
M & A Thomson Litho Ltd., East Kilbride, Scotland

CONTENTS

Birmingham violen
spreads to new areas

Thatcher calls for hel
of community leader

Police officer killed in rio

Political tensions surface a
council leaders blame polic

**Drugs and poor policing
blamed for violence**

Heroin raid in riot area

Hundreds of police
battle with mobs
in Birmingham riot

**Black jobs
gloom
worsening**

Armed youths
on rampage
in Tottenham

1   ISSUES AND CONTROVERSIES

By the late 1960s it was clear that many of Britain's inner cities were areas of crises. They were seen to be suffering from serious social, economic and environmental problems. Twenty years on, despite sets of wide-ranging policies and billions of pounds of Government money, the inner cities show little sign of improvement and conditions in general have continued to get worse. Unemployment, social deprivation and poverty have become more acute, social inequalities and racial tension have increased and in many places, the physical environment has continued to degenerate. Most recently, the effects of economic restructuring and deindustrialisation have been to deepen these difficulties to such an extent that

inner city problems are seldom out of the news. What is not often reported is that four million people have to endure the misery of everyday life in Britain's inner cities.

Although the inner city problem has become a pronounced feature of most large British cities there is considerable controversy over its *causes and the best solutions*. In Part I we shall consider the nature of the inner city problem and examine the history of Government policy towards these areas. We will discuss different explanations of the crises facing the inner city and show how policy solutions reflect these competing viewpoints.

2   DEFINING THE INNER CITY

In many respects the inner city is a difficult area to define. This is reflected in a lack of consistent official data. Different sources use varying definitions and terms to delimit the inner city, which is a major problem when trying to compare places, especially over time. Geographically, the area is inexact. In most cases the inner city refers to those areas of older (mainly nineteenth century) residential and industrial development lying between the Central Business District and the suburbs of major conurbations. To this effect, the inner city broadly corresponds to the *zone of transition* in Burgess' model of urban land-use. However, the inner city is seldom a continuous area nor exclusively central in location: its form and extent varies from place to place (Figure 1.1).

In practice, the term inner city implies much

more than a location. Peter Hall (1981) suggests that three features outline the nature of the inner city and what has become to be known as the *inner city problem*.

The first is *decline*. The inner city is an area of absolute and relative loss of both population and employment opportunities.

The second, is *deprivation*. Within the inner city are areas of poor environmental condition, large proportions of run down and deteriorating housing, high rates of overcrowding, high levels of low-skilled workers, acute unemployment and many other characteristics of poverty and social malaise.

Thirdly, these two features combine to produce a feeling of helplessness amongst many inner city dwellers. There is a *common perception of collec-*

◄ *Scenes of the inner city. The newspaper headlines date from the riots of 1985*

tive despair: a sense that individuals can no longer help themselves and little can be done to change the situation.

The inner city problem is not one problem but a complex set of interrelated issues producing a web of decline, deprivation and despair.

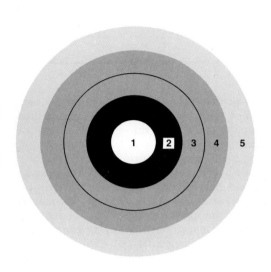

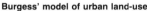

Burgess' model of urban land-use
1 = CBD
2 = Zone of transition (inner city)
3 = Zone of working men's homes
4 = Residential zone
5 = Commuter zone

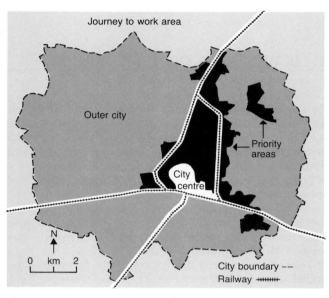

Coventry's inner city*: Priority areas (West Midlands County Structure Plan)
*Defined by 1981 census data on unemployment, overcrowding, basic amenities, lone parent households, pensioners living alone, ethnic origin, car ownership

Figure 1.1 *Defining the inner city*

3   THE NATURE OF THE INNER CITY PROBLEM

3.1 Decline

Many of Britain's inner cities are areas of *population and economic decline*.

i) Population change

Inner cities are areas of population loss (depopulation). Figure 1.2 shows that the population was moving outwards (decentralising), *relatively* during the period 1951–1961 and *absolutely* thereafter, from the *cores* of British urban areas into their surrounding *rings*. Initially the effects of out-migration were masked by high urban birth-rates, but after 1961 the rate of migration was so great that this was no longer the case. The effect of this loss varied from place to place, although all major conurbations experienced a population decline within their inner areas. Table 1.1 demonstrates the scale and extent of these changes. Inner London lost nearly one million people between 1961 and 1981 (see also Table 3.1). Four closely related reasons contributed to these patterns of population change.

First, in the aftermath of the Second World War policies of slum clearance and decentralisation to New Towns promoted out-migration from the urban cores. Both policies were attempts to relieve inner city congestion and overspill. By the mid 1970s, as a result of a massive contraction in the national house building programme, demolitions were outstripping public sector constructions in all but one of the major provincial con-

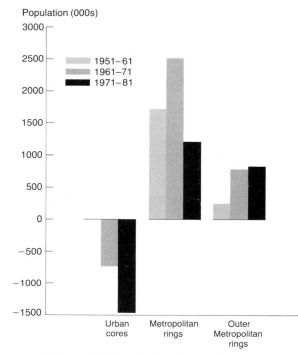

Population (000s)

1951–61
1961–71
1971–81

Urban cores | Metropolitan rings | Outer Metropolitan rings

Figure 1.2 Population change, 1951–81

Table 1.1 Population change within inner areas*

	Population change (%)	
	1961–71	1971–81
London	−13.5	−17.6
Birmingham	− 8.3	−17.6
Liverpool	−18.3	−26.6
Manchester	−16.8	−24.5
Newcastle	−17.8	−18.0

*Inner areas are the wards nearest the city centre and containing approximately one third of the city's population.

Source: Census of Population.

urbations. Also most inner city redevelopment was at a lower housing density.

Secondly, improvements to public and private transportation enabled urban workers to live further away from their place of work and, in consequence, commuting distances increased. In 1961 only 30 per cent of households owned cars, by 1981 this percentage had more than doubled.

Thirdly, the suburban attractions of a house with a garden, in a pleasant neighbourhood with access to open countryside provided positive *pulls* to an increasingly mobile population, especially when compared to the congested often polluted and physically run down environments of the inner city. These changes coincided with a growing desire for owner occupation. As standards of living improved more families were able to purchase a house of their own in locations of their own choosing. The predominantly dilapidated nineteenth century structures of the inner city provided poor housing opportunities to those without considerable capital to invest. On the other hand, the overwhelming supply of new and relatively cheap housing was found on the edge of the city, where the lower land values had attracted property developers.

Fourthly, changes in employment opportunities. We shall see that the inner city was an area of spectacular job loss, whereas the outer city provided a seedbed for new growth. During the period 1961–71 inner Liverpool lost 34 000 jobs, inner Glasgow 60 000, inner Manchester 84 000 and inner London 243 000. Whether these job losses caused decentralisation or whether they were an effect of it is difficult to judge, whatever, the results on the inner city were drastic.

The loss of population in itself need not cause problems. However this migration was highly selective. Those who moved away were the young, the more skilled and the most enterprising, leaving behind a pool of ageing, low-skilled workers who faced increasingly difficult prospects of obtaining a reasonable standard of living.

Not all migration was away from urban centres. From the early 1950s until the early 1970s those who remained in the inner city were joined by a swelling flow of new arrivals from the New Commonwealth, Pakistan and parts of Africa. Although forming only a small percentage of the United Kingdom's population, little more than 3 per cent in 1981, these ethnic groups formed significant concentrations within inner city areas. For example, in Bradford in 1971 ethnic minorities made up more than 40 per cent of the population of the two inner city wards, compared to less than 1 per cent in suburban locations. As a *replacement population*, these ethnic groups had moved into jobs and houses abandoned by whites. They formed a vulnerable and exploitable pool of cheap labour; a seedbed of social unrest had been sown.

ii) Economic change

The consequences of economic change, if anything, have been more far-reaching. Inner cities are areas of industrial decline, job loss and employment change. Plant closures and contractions are typical. In many inner cities manufacturing has virtually collapsed. Table 1.2 shows the shifting pattern of employment in recent decades. Although job opportunities in Britain as a whole have fallen the inner cities have borne the brunt of this decline, experiencing a 45 per cent drop in employment between 1951 and 1981. For a time during the 1950s the inner cities marginally increased their employment, but by the 1960s these areas were rapidly contracting as economic crisis was followed by a relentless shedding of labour.

Table 1.2 Employment change within inner areas of major cities

| | | Employment change (%) | |
		Inner areas	Britain
Total employment	1951–61	+ 1.0	+ 7.0
	1961–71	−14.8	− 1.3
	1971–81	−14.6	− 2.7
Manufacturing	1951–61	− 8.0	+ 5.0
	1961–71	−26.1	− 3.9
	1971–81	−36.8	−24.5
Private services	1951–61	+11.0	+14.0
	1961–71	−15.3	+ 4.2
	1971–81	− 6.4	+14.4
Public services	1951–61	+ 1.0	+ 6.0
	1961–71	+ 2.0	+14.5
	1971–81	− 7.4	+ 7.7

Source: ESRC, 1985.

In greater detail, three significant *sectoral* changes are evident.

First, since the 1950s the inner cities have been losing all kinds of manufacturing employment. One million manufacturing jobs were lost between 1951 and 1981.

Secondly, gains in the service sector, particularly in banking, finance, insurance and public administration in the 1970s, failed to compensate for these losses. Furthermore, these increases in services took place at higher managerial levels, whilst routine clerical jobs declined. In consequence, many of the new professional jobs were filled by commuters from outside the inner cities themselves. In 1981 some 39 per cent of inner city jobs employed commuters.

Thirdly, another significant trend in the inner cities has been the growth of the female workforce. During the 1960s whilst the inner cities were shedding male labour in manufacturing jobs, women were recruited into the service sector. Up until 1978 the great bulk of any employment increase was jobs for women. However, in many cases these were part-time and low paid. The most lucrative opportunities went to men in the outer city. As a result of these profound changes the prospects of those male workers left behind in the inner city were dire. In most cases the plants and firms which survived were dominated by older, less competitive and less innovative enterprises, demanding low-level skills highly prone to be replaced by new technology. In these circumstances few of the displaced workers moved for without large-scale retraining schemes they found it difficult to gain employment elsewhere.

The geographical changes described above are part and parcel of *the decline and restructuring* of the British economy. The inner cities have been losing manufacturing jobs, not particularly because of their locational characteristics, but because within each sector they had the oldest and least profitable industrial capacity. As a result they have become areas of disinvestment, characterised by massive outflows of capital. Physical obsolescence in the form of ageing and deteriorating buildings, outmoded equipment and poor infrastructure, especially inadequate roads, have been further deterrents to new capital. In essence, economic restructuring has meant that the inner cities have lost their locational appeal for new investment.

3.2 Deprivation

Inner cities suffer from *physical decay and multiple, adverse social conditions.*

i) Physical (built) environment

Most inner cities comprise twentieth century redevelopment and the surviving parts of the nineteenth century industrial city. Both environments are associated with their own particular problems. On the one hand, massive redevelopment in the 1950s and 1960s spawned high-rise estates which quickly fell into disrepair. Often these tower blocks were hastily constructed with poor building materials. Furthermore, they created difficult living environments, especially for the elderly and for families with young children. Today, many of these tower blocks are in advanced stages of physical deterioration characterised by broken windows, crumbling brickwork, piles of litter, graffiti, broken lifts and dirty and unkempt passageways. On the other hand, the nineteenth century city has left a legacy of buildings and infrastructure unsuitable for modern needs and which are expensive to renew. In inner Liverpool, for example, 80 000 houses were built before 1919. Typical street scenes in the inner city show boarded-up vacant lots and derelict land, old terraced buildings and shops in poor condition, exposed gable ends left over by demolition, temporary buildings, on-street parking and few people (Figure 1.3). Poor housing is a major problem. For example, inner London has 30 per cent more households without exclusive use of all amenities (i.e. bath, w.c., hot and cold water) than its outer areas. Housing prices remain low in comparison to elsewhere in the city. Residential overcrowding is high, notably in inner Glasgow and inner London. Many properties are privately rented, which is an additional difficulty simply because poorer housing conditions are commonly experienced in this sector. More than one-half of inner London's households live in private rented accommodation.

Figure 1.3 *A typical street scene in the inner city*

ii) Social environment

Inner cities are areas of poverty and deprivation. Analysis of 1981 census data for the twenty biggest cities (populations more than 190 000) in England and Wales reveals that social conditions worsen in an almost straight line from the outer to the inner city. Figure 1.4 shows some striking contrasts between the inner and outer areas of a selection of these cities, using a variety of social indicators. These contrasts generally apply to all twenty cities. With about 7 per cent of the British population the inner cities contain:

14 per cent of the unskilled workers;

20 per cent of households in housing stress;

33 per cent of the New Commonwealth and Pakistani population;

51 per cent above the national average rates of unemployment;

twice the national average of single parent families;

three times the national rate of long-term unemployment;

four times the degree of domestic overcrowding;

up to ten times the national proportion of people living below the Supplementary Benefit (Income Support) poverty line;

most of those schools classed as *exceptionally difficult* (high proportion of children from low income households and greatest concentration of children with learning and language problems).

Not all inner city residents are poor, nor are all the poor resident in the inner city. However, two examples, at different scales, illustrate the plight of these areas. In the first, a large number of urban areas were ranked according to sets of social indicators which measure disadvantage. The 10 most deprived areas are shown in Table 1.3 (in which the lower the index score the worse are the social conditions). It is clear that the list is dominated by inner city areas.

The second study focuses on Birmingham. In Figure 1.5 the wards of Birmingham have been scaled from worst to best according to a set of social indicators which inform about levels of deprivation. The inner wards of Nechells, Sparkbrook, Handsworth and Soho stand out as an area of extreme disadvantage and poverty.

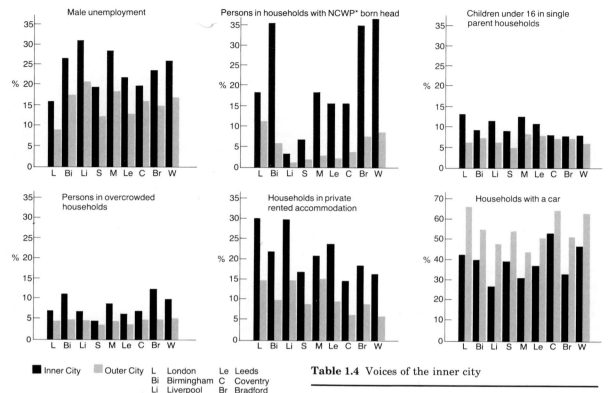

Inner City ■   Outer City □

L	London	Le	Leeds
Bi	Birmingham	C	Coventry
Li	Liverpool	Br	Bradford
S	Sheffield	W	Wolverhampton
M	Manchester		

*NCWP New Commonwealth or Pakistani

Figure 1.4 *Contrasts between the inner city and outer city, 1981*

Table 1.3 Indices of social deprivation: the 10 most deprived urban areas

Urban area	Deprivation score
Inner Birmingham	17
Inner Glasgow	22
Liverpool	23
Inner Salford	29
Glasgow new peripheral	44
Teeside	47
London Docklands Special Area	49
Inner West Midlands	49
Inner Tyne and Wear	57
Inner Leeds and Bradford	59

The list is based on 16 indicators from the 1981 Census measuring lack of choice and lack of opportunity. The areas are ranked from most to least deprived.

Based on Eversley and Begg, 1986.

Table 1.4 Voices of the inner city

Young single parent, White, female: 'There's nothing here. I look around and all I see are troubles and headaches ... My roof's leaking ... its been leaking for over three years. My landlords refused to mend it ... I keep complaining ... The kitchen's full of damp ... The walls are black. All the paper is peeling ... The mould and damp are getting at my children ... My youngest has been off school five weeks since Christmas ...'

Middle-aged, Black male: I am a tradesman ... lots of people like myself learnt their skills years ago ... But I never use them ... there's nowhere to go. I've been out of work for more than five years ... Any new developments that takes place are filled by people from outside the city, not by people from within this area. Its depressing ... why should they get our jobs?

Young Black, Male: I was chatting to my friends like (on a street) ... We were having a good time ... playing a little music ... having a few drinks ... when this rozzer (policeman) comes up. He tells us to move on ... We says we ain't doing nothing ... Then some pushing started ... Soon the street was crawling with them (policemen). Two of me mates were picked up. We weren't doing anything ... I left school last year ... What hope have I got?

Source: BBC Radio Report on Handsworth, inner Birmingham, March 1989

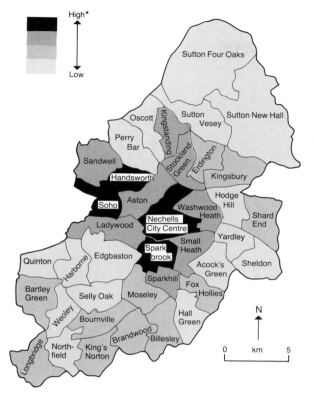

High*

Low

Sutton Four Oaks

Oscott
Kingstanding
Sutton
Vesey
Sutton New Hall
Perry
Bar
Stockland
Green
Erdington
Sandwell
Handsworth
Kingsbury
Soho
Aston
Hodge
Hill
Washwood
Heath
Ladywood
Nechells
City Centre
Shard
End
Yardley
Spark
brook
Small
Heath
Quinton
Harborne
Edgbaston
Acock's
Green
Sheldon
Sparkhill
Bartley
Green
Fox
Hollies
Selly Oak
Moseley
Weoley
Bournville
Hall
Green
Longbridge
North-
field
King's
Norton
Brandwood
Billesley

N

0   km   5

*Based on

i)   Male unemployment
ii) Persons in household with New Commonwealth or Pakistani head
iii) Overcrowded households
iv) Social classes iv and v
v)   Young, single person households

Figure 1.5 *Social deprivation in Birmingham, 1981*

In many cases inner city residents are surrounded by the signs of severe environmental blight. From the late 1950s until the early 1970s large cities shared an enthusiasm for the concept of comprehensive redevelopment. A vast slum clearance programme went ahead. In Manchester, for example, 83 000 dwellings were demolished in this period. However, clearance was so extensive and rapid that redevelopment could not keep pace with demolition and vast tracts of vacant land were left behind. In the late 1970s up to 12 per cent of inner Glasgow comprised of vacant land, 10 per cent of inner Liverpool, 7 per cent of inner Birmingham and 5.5 per cent of inner London. Unkempt, overgrown and rubbish strewn, these areas of urban wasteland have become depressingly familiar features of the inner city, adding to the impression that these areas are beyond help.

During the 1980s the inner cities have become places of growing social unrest, racial tension and rising crime. For years there had been warnings that the inner cities were living on a short fuse that could start an explosion at any time. In 1980 in the St Paul's area of Bristol serious clashes took place between the police and the local community. In the summer of 1981 the explosion occurred when several of Britain's largest cities were hit by riots. Since then major disturbances have broken out sporadically elsewhere (Table 1.5 and see p.4).

Inner cities are often areas of rejection with poor public images. The reasons for these strong negative perceptions were highlighted in a recent survey in Glasgow. The most cited deal with appearances, disorder, and dereliction (Table 1.6).

3.3 Collective despair

Decline and deprivation of this kind have left their marks on the residents of the inner city. Amongst inner city dwellers there is often *a deep sense of dismay and a common feeling that all hope is lost.* Opportunities are perceived to fall well short of what is needed. Spirits are dulled by the physical decay of factories, houses, streets; the lack of jobs, good housing, and leisure and shopping facilities; the poor level of public services; and underlying all of this, is a sense that the area is running down (Table 1.4).

Table 1.5 Inner city riots in the 1980s*

Inner city	City	Date
St Paul's	Bristol	April 1980
Brixton	London	April 1981/Nov 1985
Southall	London	July 1981
Toxteth	Liverpool	July 1981
Moss Side	Manchester	July 1981
Tottenham	London	Oct 1985
Handsworth	Birmingham	July 1981/Sept 1985

*In July 1981 minor disturbances also occurred in Blackburn, Bradford, Derby, Leeds, Leicester, Wolverhampton.

Table 1.6 Rank order of inner city problems perceived by house-buyers in Glasgow

Problem	Rank order
Pavements and roads in poor states	1
Not enough things to do for teenagers	2
Drab and decayed appearance	3
Vandalism and hooliganism	4
Waste ground	5
Empty or derelict houses	6
Too much violence and crime	7
Rubbish in streets	8
Local people not caring about area	9
An area of bad reputation	10

Source: Scottish Economic Planning Department, 1977.

Low participation rates in local elections and poor involvement in community-based organisations are further signs that Britain's inner cities are environments of alienation and despair.

3.4 Summary

The nature of the inner city problem is summarised in Figure 1.6.

Figure 1.6 The inner city's web of decline, deprivation and despair

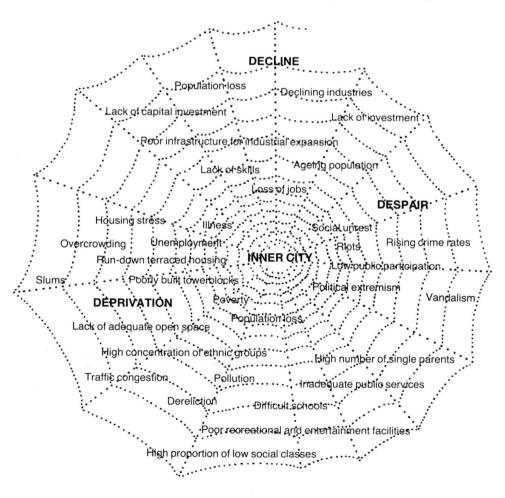

4   EXPLANATIONS OF INNER CITY PROBLEMS

Over the years explanations have differed as to the root causes of the inner city problem. Different perspectives have led to the perception of different problems and to different goals and means for tackling them. As we shall see some of these explanations have had a considerable bearing upon the content of urban policy. Six competing explanations are distinguished in Table 1.7 and these can be grouped under three headings for discussion. The reasons for these shifting ideas are given in section 5 which looks at Government policy towards the inner city over four phases, since 1946.

4.1 Explanations which focus on the physical fabric of the inner city

These views prevailed for over two decades in the aftermath of the Second World War. The problems of the inner city were attributed to run down built environments, ill-adapted structures to modern needs and poor housing (*physical obsolescence and decay*). Redevelopment and clearance were seen as solutions. Only through rebuilding (*physical planning*) would new and better environments arise (see Phase 1 below).

4.2 Explanations which focus on the social fabric of the inner city

During the late 1960s to the mid 1970s attention began to focus on issues of social deprivation. Three alternative explanations gained consider-

able support (see Phase II below). Each of these explanations led to solutions which focused upon helping particular areas or communities. Accordingly, these remedies were termed *area-based*. The inner city was not treated as a whole region, instead parts of the inner city or particular communities were targeted for special assistance.

i) Culture of poverty

This is the notion that families of a certain kind, often living in the inner city, pass on an *anti-social culture* from one generation to the next. Life-styles are encouraged which frequently lead to early school-leaving, vandalism, early marriage, early child-rearing, and crime. Such explanation suggests that deprivation is concentrated into certain areas, is essentially created by the deprived themselves, and might be solved, with little cost, through area-based social education schemes. These programmes would bring about better adjusted families.

ii) Cycle of deprivation

Many inner city families are seen caught in a *downward spiral of deprivation* (Figure 1.7).

The *cycle of deprivation* was thought to prevent the poor raising their living standards. Policies offering special assistance, in the form of improved educational facilities and enhanced community support, followed naturally from this explanation. The goal was to break the cycle and bring about better adjusted families.

Table 1.7 Different explanations of the inner city problem

Perspective	Perceived problem	Goal	Means
Physical decay	Obsolescence	Better built environment	Physical planning
Culture of poverty	Pathology of deviant groups	Better social adjustment	Social education
Cycle of deprivation	Individual inadequacy	Better families	Social work
Institutional malfunction	Planning failure	Better planning	Coordinated planning
Resource maldistribution	Inequality	Reallocation of resources	Positive discrimination
Structural conflict	Underdevelopment	Redistribution of power	Political change

Based on the Coventry CDP Report, 1975

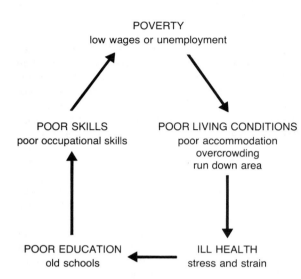

Figure 1.7 *The cycle of deprivation*

iii) Institutional malfunctioning

Problems were seen to arise from a *lack of links between populations most in need and particular services*, like community care and social and welfare services. Also, many inner cities were seen as places ravaged by poor planning. Uncleared sites of partly demolished buildings and haphazard redevelopment created areas of *planning blight*. The solution was better planning through clearer objectives and a more coordinated approach by those local agencies concerned with the regeneration process.

4.3 Explanations which focus on the structural (economic) fabric of the inner city

From the mid-1970s onwards two additional explanations were influential in shaping attitudes towards urban deprivation (see Phases II and III below). Both arose from teams working within the inner city: these were the Community Development Project (CDP) teams and the Inner Area Studies (IAS) teams. Both argued that *area-based solutions* were unlikely to solve urban problems: at best, they provided short-term help, but more often they simply displaced the problem to a neighbouring, unassisted area. What was

needed was a *total approach*, which viewed the inner city as a problem region. Quite simply, these teams saw *urban deprivation as a product of lack of income and wealth*. Particular stress was laid on the crucial issue of employment. For some time it had been clear that the inner city was an area of low wages and high unemployment, in which many people could not find work at all. It was also clear that large numbers of young, skilled people had been leaving the inner city so worsening the economic, social and physical decline of these areas. At this point the explanations of the CDP and the IAS teams differed.

i) Resource maldistribution

On the one hand, the IAS teams viewed the inner city problem as an outcome of industrial change which brought about *restructuring, reorganisation and rationalisation*. Investment in the British economy was towards newer goods and services (restructuring) and away from the traditional regions, such as the inner city (reorganisation). This massive contraction in manufacturing, coupled with a trend to concentrate production into fewer but larger units (rationalisation) located in green-field sites on the edge of cities, inevitably cut across other sectors like housing, retailing, transport provision and other services. In order to check this massive outflow of capital, resources should be allocated to favour inner city locations, in order to make them attractive locations for new enterprises oncemore.

ii) Structural conflict

On the other hand, the reports of the CDP teams, especially from the Coventry team, argued that the inner city problem arose from *a system of production* which was based on the accumulation of private profit. This report claimed that whilst large manufacturing industries bring prosperity to some areas or parts of cities this can only be maintained at the expense of other areas. Large companies benefit from keeping the inner city *underdeveloped*. These areas are seen to act as a buffer zone, providing a pool of expendable, local, unskilled labour, and cheap land, which can be used or dumped according to fluctuations in the economy. Inner city decline would be halted only if there was a vast change in economic power and political organisation: capitalism had to be replaced by a socialist system of production.

5 GOVERNMENT POLICY TOWARDS THE INNER CITY

Four broad phases of government policy towards the inner city can be distinguished. Each phase is associated with a different set of explanations and solutions to the inner city problem (see Explanations of inner city problems). At the end of Part I Figure 1.10 summarises the major policy initiatives aimed at solving inner city problems.

5.1 Phase I (1946–1967)

In the immediate post-war period inner cities were seen as places of physical decay and poor housing, most of which were in need of drastic redevelopment. As a result in 1947 the newly elected Labour Government introduced the concept of *Comprehensive Development Areas* which launched a massive programme of slum clearance and rebuilding aimed at making these inner areas less congested and environmentally more attractive. In the next twenty years some one and a half million properties were knocked down as part of this scheme.

This sort of policy was based on the belief that the problems facing these inner areas were not more than highly localised events. Most inner cities comprised of old working class housing built hastily during periods of rapid industrialisation and urbanisation in the nineteenth century. The types of housing varied from place to place: in the northern towns one room up, one room down, back-to-backs were typical, whereas in London multi-storey tenements were constructed because of the pressure on land. By the late 1940s these industrial inner areas, essentially the creation of the Victorian period, were showing signs of decline and disrepair. Buildings were overcrowded, poorly ventilated, many lacked basic amenities and most were in a run down condition. In addition, many inner cities had suffered considerable bomb damage. A quick solution was to clear these areas and start anew. *Physical planning* of this kind was seen as an effective remedy and offered a means by which new and pleasing environments would be created.

These ideas apart, inner city policy lacked any further direction. Little thought was given to the social consequences of redevelopment. In many respects Comprehensive Development Areas formed only a subsidiary consideration as the main thrust of urban policy was towards decentralisation. The New Towns Act (1946) and The Expanded Towns Act (1952) encouraged the movement of urban residents to newer, smaller and, by implication, better settlements beyond the major conurbations. Urban growth was to be further contained by the imposition of encircling green belts.

Comprehensive Development Area policy was not entirely successful. Towards the end of this period a number of problems became apparent.

First, in many cases slum clearance was so extensive that redevelopment could not keep up with demolitions. In consequence vast tracts of derelict and vacant land were left behind.

Secondly, whole communities were displaced by slum clearance. Those people that remained found themselves removed from their previous neighbours, family and friends. A mood of despondency became widespread.

Thirdly, the tower blocks which were built to replace the slums proved to be hostile and unsuitable environments, especially for the elderly and the very young. Many soon fell into disrepair and instead of being full of low-rise slums the inner city was now full of high-rise ones.

Fourthly, the policy was too limited in focus. It failed to halt the economic and social decline of the inner city. Emphasis upon decentralisation and redevelopment, at lower densities, encouraged the outward movement of young, skilled workers and accelerated the loss of manufacturing employment.

Lastly, redevelopment was failing to keep pace with decay. The House Condition Survey of 1967 showed that 1.8 million houses in England and Wales were unfit for human habitation and a further 4.5 million were in need of repair. These findings were used to argue that clearance would never be a once-and-for-all solution to urban decay: houses are continually ageing and so there is always going to be a need for replacement and repair. This thinking led to a call for a shift from wholesale clearance to rehabilitation and improvement.

5.2 Phase II (1967–1977)

This was a period of major policy change. For the first time attention began to focus on the complex order of inner city problems, which were now seen in other than physical terms. A number of factors seem to have encouraged the development of Government thinking on the inner city.

First, disenchantment with the effects of redevelopment and decentralisation upon the inner areas of the larger conurbations (see criticisms of Phase I).

Secondly, an increasing concern for the related issues of race and immigration, and multiple deprivation. Between 1951–1966 Britain's Black population increased from an estimated 74 500 to 595 000 and the vast majority of these settled in inner city areas. Often these new immigrant communities were highly concentrated into particular neighbourhoods and streets. It was increasingly obvious that many of their members faced acute problems in housing, education and employment, frequently stemming from racial discrimination. Tension developed between immigrant and host communities and it was feared that without intervention the race riots that had occurred in the USA in 1967–68 might also happen in Britain.

Thirdly, unemployment was high and rising in the inner city and this was an unfortunate feature for a Labour Government which counted the inner city as its heartland.

Fourthly, a growing body of opinion suggested that housing and environmental improvement schemes were cheaper and more effective mechanisms for urban regeneration than clearance. Rehabilitation was socially less disruptive and economically more viable. It was estimated that at least 200 000 houses would need to be replaced each year because of continuing obsolescence, whereas with relatively small financial investment both the surrounding environment and older housing could be rapidly improved.

Fifthly, American experience with inner urban policy was proving successful in combating similar features evident in British cities. British civil servants were eager to embark upon similar innovation.

In response to these issues a broad set of policies were implemented by successive Labour (1964–70: 1974–79) and Conservative (1970–74) Governments. This was *an era of experimentation*, which in part reflected an uncertainty as to the causes and best solutions of inner city malaise. Consequently, many of the schemes were *small in scale and limited in focus*, often only assisting parts of the inner city or their population. In time such a strategy was recognised as an *area-based, needs-led approach* to the problems. Two strands to inner city policy were developed.

i) Policy for environmental improvement

General Improvement Areas (1969–). In 1968 the Labour Government published the White Paper, *Old Houses into New Homes*, which led to the 1969 Housing Act. This Act was the first main attempt to shift attention away from clearance towards rehabilitation. It enabled Local Authorities to set up General Improvement Areas (GIAs) in the older parts of cities, within which grants would be available to improve both the housing and the surrounding environment (Figure 1.8).

Figure 1.8 *Thurgarton Avenue, Central Sneiton General Improvement Area, Nottingham. The houses have been improved with new windows and roofs. The street has been pedestrianised, with new paving, lighting, and landscaping*

Housing Action Areas (1974–). The 1974 Housing Act strengthened these moves towards inner city rehabilitation by establishing Housing Action Areas (HAAs). These were designed to give priority to the improvement of areas of greatest housing and social stress. Emphasis was entirely upon housing improvement and higher rate grants were available. If HAAs were successful they could be upgraded into GIAs.

The experience of inner Manchester was typical of most large cities. Some 40 000 dwellings were estimated to be suitable for rehabilitation rather than clearance and redevelopment. By the end of the 1970s 38 HAAs and 8 GIAs had been declared, with a total housing stock of over 18 000 dwellings.

Although both developments are important in that they established the principle of area-based renewal, their success was limited for two reasons. First, in some cases these renewal strategies contributed to the problems of the inner city. They did so by encouraging the movement-in of the better-off who saw their chance of buying cheap rented property for improvement and owner occupation. In this process, known as *gentrification*, many existing inner city residents were displaced only to face the problems of acute housing shortage. Secondly, in both cases designation was slow, largely because of underfunding. By 1977 some 360 000 dwellings had been improved but many more remained in need of attention.

ii) Policy for social and economic welfare

These experiments can be further sub-divided into three groups, each reflecting a different explanation of inner city problems: first, resource allocating projects; secondly, those concerned with improving the delivery and coordination of urban services; and, thirdly, investigative projects.

a) Resource allocating projects

Two major initiatives were launched both rooted in explanations like the *culture of poverty* and *the cycle of deprivation*.

The Urban Programme (1968–). This was a major attempt to tackle 'the problems of areas where immigration had been at a high rate and . . . multiple deprivation'. It was a policy of positive discrimination, which allocated resources to areas of need. The Rate Support Grant provided 75 per cent of the costs of the projects, the remainder was found by local authorities. Local Authorities and voluntary agencies were invited to bid for grants. Several thousand projects were

approved, many involved holiday schemes, pre-school provision, child care and community work. In the first ten years of the programme more than £62 million was allocated.

Educational Priority Areas (1967–). These were set up in response to the Plowden Report on *Children and their Primary Schools*. This Report was among the first to recognise the existence of areas of severe deprivation in British cities, many of which were in inner localities. The Government's response was to fund Educational Priority Areas (EPAs). This was an area-based programme, which redistributed resources to schools in problem areas. Benefits included in-service training and special payments for teachers, attached social workers, community schooling and expanded pre-school provision. Most of the schools to benefit were in the inner city, although not all inner city schools were favoured. In time the concept of EPA became diluted to EP school, with extra resources given to single schools alone.

b) Urban services projects

In some instances inner city problems were seen to arise when communication between residents and services was impaired.

Comprehensive Community Programmes (1974–1979). These programmes were introduced with the aim of improving delivery and coordination of services by local authorities and other welfare agencies in areas of intense deprivation. About 90 areas in Britain qualified for special assistance through the concentrated nature of their urban problems. The feasibility of this scheme was tested in a series of experiments at Motherwell, Gateshead, Bradford and Wandsworth. Largely underfunded, CCPs never progressed very far.

c) Investigative projects

Two important investigative projects were undertaken to look further into the causes of inner city decline; their findings had a profound effect on changing official attitudes to deprivation (see Explanations of inner city problems).

Community Development Projects (1969–1977). 'A neighbourhood-based experiment aimed at finding new ways of meeting the needs of people living in areas of high social deprivation'. Twelve projects were established, mostly in inner city areas, with a central team at the Home Office. A total budget of £5 million was allocated. Emphasis was upon improving links with all local social services and the encouragement of self-help groups. In time the CDP team clashed with their parent local authorities and the Home Office. A number of their reports, published in the mid-1970s, began to question the basis of area-based solutions to urban poverty. These reports so angered the Home Office that all financial support was withdrawn and by 1977 each team had closed.

Inner Areas Studies (1972–77). Three research teams were commissioned by the Department of the Environment to look at the problems of the inner areas of Birmingham (Small Heath), Liverpool (Vauxhall), and Lambeth (Stockwell). The reports were made in 1977 and their lasting impact was to reject many of the prevailing explanations of urban deprivation, in favour of structural (economic) solutions. The main message coming from these investigative projects was that the inner city problem would not be solved through small-scale urban experiments. Inner city malaise was seen to be the outcome of broad structural (economic) changes in the British economy. Instead, a *total approach* was needed which would treat the inner city as a problem region.

5.3 Phase III (1977–1979)

A major shift in inner city policy came in the mid-1970s. The publication of the influential White Paper, *Policy for the Inner Cities*, in 1977, brought to an end the period of urban experimentation. A Cabinet Committee had been set up in 1975 to review inner city problems. In their final report they rejected cultural and cyclical explanations of deprivation in favour of *structural (economic) arguments* (see Explanations of inner city problems). For the first time the inner city was defined as a *problem region*. Broad based action for the inner city was recommended, combining social, economic and environmental programmes. In particular, emphasis should be given to restoring the industrial base of the city from which other improvements would follow. Official recognition of these ideas quickly followed in what has become *a period of substantive policy*

towards the inner city. However, controversy surrounds the best way to revive the economic fortunes of the inner city. Although both Labour and Conservative parties claim commitment to the inner city the methods each has adopted when in government have differed. Accordingly, Phase III corresponds to the period of Labour government.

i) A total approach to inner city problems

The Enhanced Urban Programme. An important recommendation of the White Paper was the creation of *Partnership and Programme Areas* in selected inner cities. Here, central government would enter into formal agreement with local authorities and voluntary organisations to produce a coordinated strategy to assist the economic, social and environmental regeneration of the *whole inner city*. This would be funded by enhancing the resources allocated to the traditional Urban Programme and transferring the coordinating role to the Department of the Environment. These ideas were sanctioned by the *Inner Urban Areas Act* of 1978. The first partnerships were offered to London's Dockland authorities, Hackney and Islington, Lambeth, Birmingham, Liverpool, Manchester, Salford, and Newcastle and Gateshead. Separate arrangements were devised in Scotland through the Glasgow Eastern Area Renewal Project (GEAR), coordinated by the Scottish Development Agency.

A key objective of this new urban policy was to strengthen the economies of inner cities, in particular to encourage the retention of jobs. In recognition of this, Partnerships could declare *Industrial Improvement Areas* within their urban cores for which a variety of grants were available to employers for improvements and conversions. It was hoped that further *public spending* on basic services and the physical environment would attract new private investment and thereby generate a self-sustaining growth. Additional financial support for these new strategies came from tilting the Rate Support Grant to favour the inner cities. In the first two years of the Enhanced Urban Programme funding rose from £30 million to £165 million. It is interesting to note that in all of this activity, most of which stressed the importance of economic decline, no more than 28 per cent of Urban Programme funding was earmarked for economically oriented projects.

The *traditional Urban Programme* was not dismantled as a result of these changes. Any urban local authority could still apply for a grant. However, it was clear that a substantial proportion of the enhanced inner city investment would go in future to Programme and Partnership Authorities.

ii) *A complete inner city policy*

These urban programmes were seen as only a part of a broader plan towards inner city revitalisation. A wide-ranging national assault on the inner city problem was to be made. Other steps included winding down New Town and Expanded Town projects in an attempt to halt the continuing decentralisation of people and jobs. Also, all Metropolitan Authorities were encouraged to *bend* their mainstream programmes for housing, education, social services and transport towards the inner city. *All of these plans relied heavily on public sector support in combination with private investment.* However, the Labour Government was destined for only a short term in office and during this time little progress towards this national campaign was achieved.

5.4 Phase IV (1979 to date)

Most of the initiatives begun by the Labour Government of the late 1970s were continued by later Conservative governments, although many additions and modifications have subsequently been introduced. Commitment to inner city policy has recently been reaffirmed with the publication of *Action for Cities* (1988), which outlined the breadth of the Government's programme intended 'to help cities meet the challenge of change', and the appointment of a minister for inner cities. This section will describe current inner city policy and then evaluate new developments in a broad context.

i) *Contemporary urban policy*

Three principal aims have shaped the form and content of recent policy: securing economic regeneration; improving the environment; and gearing services and amenities to the needs of local communities. Of these, it is the economic aim which has been dominant. The Government claims commitment to policy which attempts to tackle the long term decline and change in the economic fortunes of the inner city by restoring confidence, initiative, enterprise and choice. Figure 1.9 shows contemporary urban policy initiatives in England and Wales. The main programmes are outlined below. These can be grouped under five headings.

a) *Urban development programmes*

The Enhanced Urban Programme. Programme assistance has been extended to cover 56 localities, including most major cities, of which 7 are still classified as Partnerships (Figure 1.9). However, the list has slightly altered since 1979 in that Docklands is no longer so classified and Islington/Hackney has been split into two Partnership authorities. The 7 Partnerships received about 47 per cent of Urban Programme funding in 1987. Greater emphasis is now placed on funding projects with an economic focus. Throughout the last decade spending has tended towards a 40/40/20 split between economic, social and environmental schemes. Over 12 000 projects were running in the late 1980s with funding amounting to about £360 million. In 1986 the *traditional Urban Programme* was discontinued.

Urban Development Corporations (UDCs). The *Action for Cities* document describes UDCs as 'the most important attack ever made on urban decay'. In effect, these are large, single-purpose, Government sponsored agencies which were introduced in 1981 with the sole objective of regenerating large tracts of derelict land in the inner city. To achieve this, a UDC seeks: 'to bring land and buildings into effective use; to encourage the development of industry and commerce; to ensure housing and social facilities are available to encourage people to live and work in its area'. The first UDCs were established in London's Docklands (see Part II) and on Merseyside. In 1987 five additional UDCs were created in the Black Country, Cardiff Bay, Teeside, Tyne and Wear and Trafford Park (Manchester), with three smaller agencies for Bristol, Leeds and Manchester. In the following year the Lower Don Valley UDC was declared. There are apparently to be no more UDCs for the forseeable future. In 1988/89 UDCs managed over 100 000 hectares and received £200 million in grants. The role of UDCs has been likened to that of New Town Corporations and accordingly, this programme has been described as *In-Town New-Town* policy.

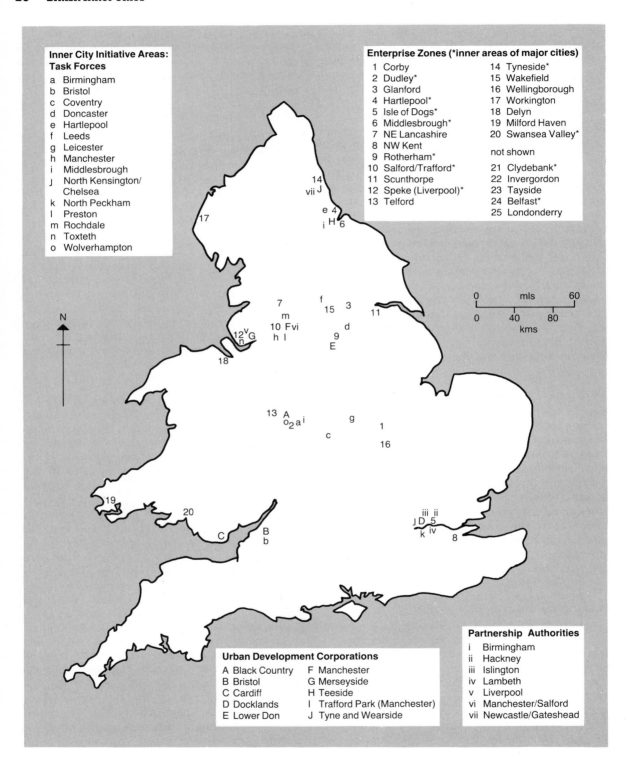

Inner City Initiative Areas: Task Forces

a   Birmingham
b   Bristol
c   Coventry
d   Doncaster
e   Hartlepool
f   Leeds
g   Leicester
h   Manchester
i   Middlesbrough
J   North Kensington/ Chelsea
k   North Peckham
l   Preston
m   Rochdale
n   Toxteth
o   Wolverhampton

Enterprise Zones (*inner areas of major cities)

1   Corby
2   Dudley*
3   Glanford
4   Hartlepool*
5   Isle of Dogs*
6   Middlesbrough*
7   NE Lancashire
8   NW Kent
9   Rotherham*
10 Salford/Trafford*
11 Scunthorpe
12 Speke (Liverpool)*
13 Telford

14 Tyneside*
15 Wakefield
16 Wellingborough
17 Workington
18 Delyn
19 Milford Haven
20 Swansea Valley*

not shown

21 Clydebank*
22 Invergordon
23 Tayside
24 Belfast*
25 Londonderry

Urban Development Corporations

A Black Country
B Bristol
C Cardiff
D Docklands
E Lower Don
F Manchester
G Merseyside
H Teeside
I Trafford Park (Manchester)
J Tyne and Wearside

Partnership Authorities

i   Birmingham
ii   Hackney
iii Islington
iv   Lambeth
v   Liverpool
vi   Manchester/Salford
vii Newcastle/Gateshead

b) Deregulation and private sector initiatives

Enterprise Zones (EZs). Introduced in 1980, these were a major attempt to stimulate economic activity by lifting certain financial burdens and administrative controls. They have a ten year life-span during which the principal benefits are exemption from rates, tax allowances for capital expenditure on industrial and commercial property, and a simplified planning regime. Twenty five Enterprise Zones have been set up throughout the United Kingdom, although all of these are not within the inner city (Figure 1.9). They vary in size from 50 hectares (Glanford) to over 450 hectares (Tyneside). By 1986 it was estimated that 63 000 jobs were located in the zones, of these two-thirds were in manufacturing, one-quarter were start-ups and about one-third were direct transfers into the zones. Between 1981–86 £300 million was spent and 700 hectares of inner city land developed.

City Action Teams (CATs). These were set up in 1985 in London, Birmingham, Liverpool, Manchester and Salford, and Newcastle and Gateshead, and in 1987 in Leeds and Nottingham. CATs are teams of civil servants who encourage joint action between Regional Government Departments in order to stimulate new private sector investment in the inner city.

The Inner Cities Initiative (ICI). Launched in 1986, this initiative created *Task Forces* in selected smaller areas where unemployment was severe and help was needed by the local business community. Task Forces are located in 17 inner city areas (Figure 1.9) to try out new ideas and approaches in order to 'unlock development opportunities'. The most recent addition is the Toxteth–Granby (Liverpool) Task Force (March, 1989) set up as a result of the Government's annual review of its *Action for Cities* programme.

c) Regeneration and redevelopment programmes

City Grants. As part of *Action for Cities* a new City Grant has been implemented. This replaces and combines three previous schemes: *the Urban*

Development Grant* (UDG, 1982), *the Urban Regeneration Grant* (URG, 1987) and *the Derelict Land Grant* (DLG, 1983). All of these shared the objective of trying to lever private sector investment into the inner city, by developing run down areas (UDG), by redeveloping large industrial sites and refurbishing large groups of buildings (URG) or by replanning derelict land (DLG). The new, streamlined City Grant helps to remove the overlap of the different grant regimes and thereby reduce uncertainty. Priority is given to the 56 Programme localities, although local authorities are *not* involved in devising and assessing applications.

Land Registers. It is likely that the City Grant scheme will run hand-in-hand with the use of Land Registers. Since 1981 vacant land held by local authorities has been identified through Land Registers and offered for sale to private developers. About 125 000 hectares have been identified, sold and bought through this process, although there is still some 220 000 hectares available for development.

d) Training, education and employment projects

The Government has introduced a broad range of programmes aimed at training and enterprise, many of which are targeted at the inner city. These include: *Youth Training Scheme, Community Programme, Restart Programme, Local Enterprise Agencies, Small Firms Advice, Enterprise Allowance Schemes, Adult Training.* A special initiative is *The Action for Jobs* campaign, based on partnership between the Manpower Services Commission (MSC) and the private sector. The MSC has made the inner city a priority for its employment resources with its rapidly expanding budget which now stands at £3 billion. In terms of training these programmes involved almost half a million inner city residents in 1987. Another initiative includes the setting up of *City Technology Colleges* in Middlesbrough, London, Birmingham and Nottingham.

e) Special projects

As part of *Action for Cities* a number of other projects were set up in the inner city. These include: the *Safer Cities* scheme, a project designed to tackle crime and fear of crime in 20 inner city areas; arts developments, like the establishment of the Merseyside Maritime Museum; and the

◄ *Figure 1.9* Urban policy initiatives in England and Wales, as at April 1989

British Urban Development Company (BUD) a consortium of 11 engineering and construction companies intent on developing inner city sites.

ii) New directions in urban policy

Taken as a whole, these initiatives represent *three* fundamental shifts in Government thinking towards urban policy. In summary, these are a movement from *local to central control*, from *public to private sector involvement*, from *social to economic concern*.

Local to central control. The scope of central Government involvement in the day to day administration of urban policy has considerably increased since 1979. At the same time the control of local authorities has been lessened. A number of reasons prompted this change:

first, Conservative Governments argue that local bureaucracies and red-tape delay the speedy up-take of new schemes and hold back business initiatives;

secondly, public sector planning controls were seen to inhibit the growth of industry in the inner city, especially in the case of small firms, through the imposition of stringent regulations which regulated the type, form, and size of new developments;

thirdly, 54 of the 56 Partnership and Programme authorities are in Labour controlled inner city areas. In order to ensure that central, Conservative Government, initiatives are successfully implemented local councils have been given less chance to interfere.

Six Government Departments are now involved: Trade and Industry, Education and Science, Employment, Environment, Transport and the Home Office. Some examples of central control include: developers have to apply direct to central Government for funding through the City Grant scheme; UDCs, which are elected by central Government and locally non-accountable, control all aspects of urban development in their defined areas and have frequently reversed the decisions of local councils; employment and training schemes are closely monitored and financed by the Department of Employment in conjunction with the Manpower Services Commission.

Public to private sector involvement. Local authorities have lost further power and influence through the commitment of the *New Right* Conservative Government to the ideals of a free-market economy, based upon minimum local intervention and maximum opportunity for private enterprise. The role of public sector organisations have been consistently reduced by the Conservatives and in their place enterprise culture has been duly brought to the inner city. *Market-led* or *demand-led* strategies are encouraged, whereby private investors are left to develop what they consider to be profitable land-uses. A striking development has been the imposition of new funding arrangements. Since 1979 the Rate Support Grant, which the Labour Government had envisaged as the main source of finance for its urban policy, was tilted away from the inner city. Instead, inner city projects were to be specifically targeted through a broad set of new grants. In this way finance could go direct to local developers rather than local councils. The growth of EZs, CATs, Task Forces and the City Grant scheme all reflect a strong commitment to private enterprise and the increasing marginalisation of public sector organisations.

Social to economic concern. Since 1981 economic issues have become dominant. Concern for the creation of new jobs and training have swamped the initial worries about social and community problems. The roots of the Urban Programme of the late 1960s had lain in social concerns about deprivation and ethnic disadvantage. The early years of the Enhanced Programme still gave considerable weight to social and community issues and provided an opportunity for voluntary organisations to play an important role. However, new Department of Employment guidelines in 1981 helped to swing concern and expenditure towards attempts at job generation, the provision of new premises and skills training. Despite the urban riots of 1981 and 1985 most initiatives continued to have a strong economic bias. By encouraging private companies to invest in the inner city social benefits were thought to accrue. Between 1979/80 and 1988/9 spending on social and community projects fell from £197 million to £118 million. Not only has social concern been reduced it has largely been disconnected from mainline Government policy. It is now mainly up to voluntary organisations to claim for and administer social actions within the inner city.

iii) Criticisms of current inner city policy

A number of criticisms of current inner city policy have been made.

First, *insufficient funding*. Resources to cities have either stood still or been reduced. The extra funding which has been specifically targeted at cities needs to be balanced against the declining income of urban local authorities. Compared with 1981/2 the Rate Support Grant to the 56 Urban Programme cities was reduced by some £143 million in 1984/5. Over the same period these same cities gained only £127 million from the Urban Programme itself. Furthermore, the formal spending on urban policy has been relatively small when set against the massive expenditure elsewhere in the economy, like on Defence and Research and Development, neither of which has an urban nor regional focus. During the mid-1970s defence procurement exceeded £8500 million. Indeed, investment of this kind has encouraged considerable growth outside of the inner city and has been used to call into question the Government's commitment to urban regeneration.

A similar observation could be made about the Government's industrial policy which, since 1979, has done little to improve the position of inner cities. Most forms of direct state support for industry have been cut-back. Particularly affected have been the older manufacturing industries like steel, shipbuilding, and motor vehicles. Historically, these industries have employed considerable numbers of inner city workers. One of the few fields in which the Government has offered financial support is in advanced technology, where a number of schemes have been set-up to encourage the development and application of new processes. For a variety of reasons, including better environment, lower operating costs, and less unionised labour, the newer industries have shied away from the inner cities in favour of smaller towns and suburbs. This is the case of the electronics industry in Scotland, which has kept away from the cities and settled either in New Towns like East Kilbride, Glenrothes and Livingstone, or in small towns in the Central Lowlands. Disinvestment of this kind in the inner cities has led to claims that *the Government needs to be much more aware of the locational implications of its national policies*.

Secondly, *renewed public sector involvement is needed*. Local councils are increasingly being shut out of policy initiatives. The Government's view is that inner city regeneration is to be *private sector-led*. Both the BUDC, the newly set up business consortium, and the CBI, in its report on the inner cities (1988), argue that public sector investment is needed in order to make many sites attractive for private development. An estimated £50 billion needs to be spent to clear the derelict sites of the inner city. This sum is unlikely to be forthcoming from private developers, especially when they can locate elsewhere at a lower cost.

Also, local authorities are more likely to be sensitive to local issues and the needs of their citizens. A major criticism of UDCs, like the London Docklands Development Corporation, is that they behave rather like *plantation authorities*, wielding neo-colonial like powers to the advantage of external private developers and not the local population. The mismatch between new jobs and local skills seems an almost inevitable consequence (see Part II).

Thirdly, *the approach is too fragmentary, and lacks a clear focus*. The *total approach* advocated by the 1977 White Paper has been dismissed. Rather than a central strategy under a single ministry, no less than six Government Departments are involved, as well as numerous voluntary organisations and many private developers. Clearly, with so many different players and policies confusion of responsibilities and conflicting goals are bound to result.

Fourthly, *the consistent emphasis on helping small areas appears unsuccessful*. There is now a long tradition of this type of response, dating back to the EPAs and the CDPs of the late 1960s and continued today through UPs and EZs. The intention is to channel resources to the worst affected parts of the city in the hope that general improvement will follow. Such an approach may have some justification where problems are the result of the physical fabric of the locality. However, where the problems are more deep-rooted the strategy has major limitations. Arbitrary boundaries drawn around particular areas do not isolate them from the wider changes taking place at the core of the urban economy, such as industrial restructuring. At best, this type of response simply displaces the problem to elsewhere. In other words, the development of an assisted area may lead to problems in surrounding areas. The additional resources allocated to targeted areas often have been diverted at a direct cost to other areas. In consequence, the result of this redistribution has been to encourage division and competition between communities for jobs and services and create an adverse *border effect*.

Fifthly, *Government policy is simply not doing enough for most inner cities*. Apart from the London Docklands, there is little sign elsewhere that the problems of the inner city are waning. Of the three main aims of Government policy (see p.19) both the economic and social remain largely unfulfilled, only in terms of environmental improvement can any large scale achievement be claimed.

On the economic front it is difficult to be very positive. A recent review of Urban Programme spending estimated that it took between £7000 and £18 000 to create a new job and that for £137 million spent in 1981–84 only some 5000 real jobs were created. This figure needs to be set against the loss of 38 000 jobs in these Programme areas over the same period. Furthermore, spatial targeting has been ineffective. The open nature of the job market means that the jobs created have not benefited those who live in the inner city. Nationally, almost 40 per cent of new jobs in inner city areas are filled by commuters from outside. In some places the percentage is much higher. For example, a recent survey of inner Newcastle-upon-Tyne revealed that of all employees in new establishments only 37 per cent were recruited from inner area residents, most commuted-in each day.

On the social front too, wide ranging problems are still evident. Emphasis on economic development has often been at the expense of a more balanced concern for community development and in consequence it is on the human side that Government policy is lacking. Worsening employment and housing conditions have produced tensions between ethnic groups and the host communities. In 1981 the *Scarman Report*, which had been set up to consider the reasons for the social unrest in many British cities in that year, emphasised the need for a national policy to combat racial disadvantage. Some nine years on, little has been attempted and racial harassment has grown progressively worse. A glimmer of hope is that some ethnic groups, after long having stood aside, are beginning to become involved in inner

city initiatives largely through grants and resources made available to voluntary agencies through the Urban Programme. Also, an ethnic dimension has been added to the work of the Task Forces, two of which are headed by Blacks, and through specific schemes like the recently announced *Ethnic Minorities Business Initiative*.

It is easier to see success in terms of environmental change. Inner Manchester provides an example. A number of developments have seen the rehabilitation of its central core: the abandoned Central Station has been redeveloped for £24 million as a major exhibition centre, G-Mex; nearby Castlefield now forms a major tourist attraction based on the conversion of the world's first passenger railway station into a museum alongside a Roman fort; a third area, Smithfield, on the site of abandoned markets, has been converted into a craft and garden centre, together with a new council estate and a range of small businesses; elsewhere private housing, both in terms of newbuilds and conversions, have changed the character of former problem estates within the inner city. All of this activity and more, has brought a sense of things happening within the inner city, which had long been absent.

These environmental changes need to be balanced against the enormity of inner city malaise. The plight of inner Glasgow highlights the weaknesses of current programmes. In 1987 some 44 500 dwellings in the inner city (15 per cent of the total housing stock) were below the *tolerable standard*. Some 80 000 dwellings were in need of repair at a likely cost of more than £1.3 billion. Unemployment was high and rising. In the inner areas of Woodlands and Easterhouse it exceeded 42 per cent and 37 per cent respectively. Over 50 per cent of the unemployed had been out of work for one year or longer. At present there is little clear evidence of *general and long-lasting improvements* in the economic, social and environmental conditions of the inner city.

5.5 Summary

Figure 1.10 provides a summary of the major policy initiatives undertaken by different Governments in their attempts to curb the inner city problem. The outcome of more than forty years of Government activity is generally disappointing. Only in recent years has there been any real

Date	Policy Initiatives	Policy Phase	Government	Policy Shift

PHASE I

Date	Policy Initiatives	Policy Phase	Government	
1946	New Towns Act	Redevelopment	Labour (Attlee)	Decentral-isation ↕ Inner City Physical planning ↕ Social planning
1947	Comprehensive Development Areas			
1952	Expanded Town Scheme		Conservative (Churchill/Eden/ MacMillan/ Home)	
1964	Deprivation Studies	Discovery	Labour (Wilson)	

PHASE II

Social Local Public Needs-led Area-based

1967	Educational Priority Areas / *Old houses into New Homes*			
1968	Urban Programme	Urban Experiments:		
1969	General Improvement Areas/ Community Development Projects			
1970		Regeneration and Rehabilitation	Conservative (Heath)	
1971				
1972				
1973				
1974	Housing Action Areas/ Comprehensive Community Programmes		Labour (Wilson/ Callaghan)	
1975				
1976				

PHASE III

1977	*Policy for the Inner Cities*			
1978	Inner Urban Areas Act: Partnership and Programme Areas/Industrial Improvement Areas	Enhanced Urban Programme		Total approach

PHASE IV

1979			Conservative (Thatcher)	
1980	Enterprise Zones			
1981	Urban Development Corporations /Land Registers	Inner City Enterprise		
1982	Urban Development Grants/ Training, education, employment projects			
1983	Derelict Land Grant			
1984				
1985	City Action Teams			
1986	Inner Cities Initiative: Task Forces			
1987	Urban Regeneration Grant			
1988	*Action for Cities*: Minister for Inner Cities/City Grants/British Urban Development Company			
1989	Annual Review of inner city policy			

Eco-nomic Cen-tral Pri-vate Market-led Area-based

Figure 1.10 *Inner city policies, 1946–89*

sense of change. A consensus seems to be emerging that economic growth is the precondition of action on all fronts. If economic problems can be tackled their solution can form the context for tackling social and environmental problems.

Anthony Downs, an American economist and keen observer of British urban policy, argues that inner city initiatives have been flawed by two characteristics. The first he calls *the law of inescapable discontinuity*, whereby urban policy seems afflicted by a *stop-go syndrome*. Successive Governments appear reluctant to continue with earlier initiatives, especially if associated with an opposing political party. Accordingly, there is little continuity in inner city policy and funding has been turned off and on at political whim. The second he terms *the law of compulsive innovation*, whereby newly elected Governments launch into new schemes, often without sufficient consultation, for the sake of setting up something new.

6   FUTURE RESPONSES TO THE INNER CITY PROBLEM

Debate continues to surround how Governments should respond to the inner city problem. Clark (1989) identifies five different strategies.

The first is to continue with present *revivalist policies*, in the hope of reversing decline and promoting urban renaissance.

The second approach recognises that current revivalist strategies cannot possibly succeed. Present policy does not do nearly enough and is relatively powerless to halt the massive outflows of capital, population and jobs from the inner city. If policies of revival are to have any chance of success then planners need to be given more wide-ranging powers. A *high interventionist strategy* would feature 'restrictions on residential developments outside cities so as to inhibit urban to rural migration, massive investment in urban infrastructure, especially roads and . . . industry. Green Belts would be extended outwards . . . industry in the cities would be subsidised out of the public purse on account of its value as a generator of employment. Inner areas would benefit from positive discrimination in the allocation of resources for education and social services'.

The third strategy recognises that so strong are the shifts of population and industry which are currently taking place in Britain, planning policy is unlikely to reverse them. A solution would be to expand the boundaries of cities in order to *recapture* the growth which is currently taking place at their edges. *Redistricting* of this kind would lead to lost revenue being retaken and reinvested elsewhere, such as the inner city, in order to restore locational appeal.

The fourth strategy is based on the belief that the inner city problem cannot be reversed by local planning actions, whether public or private sector led. Instead, broad *structural policies* are the only solution to inner city decline. Despite more than two decades of revivalist policies the inner city problem has not gone away. The problems of the inner city need to be seen in a much wider context. Many of the difficulties experienced by the inner city reflect deep-rooted structural changes which have taken place at the core of the British economy. Deindustrialisation and job losses are seen as the causes of economic malaise, unemployment and social tensions. In consequence, a return to economic growth is advocated as the most effective means of restoring affluence.

The fifth strategy is to *encourage inner city decline*. Very simply, the decline of the inner city has not been halted because its decline is irreversible. Cities are no longer appealing locations for a wide range of industrial, retail, commercial and residential investors. However, although there are negative aspects, many elements of decline could be used to the positive advantage of the inner city. For example, depopulation enables the inner city to be remade at lower densities, so reducing congestion and overcrowding. Better use could be made of the built environment: old housing and industry could be cleared, roads improved and the inner cities greened as parkland and recreational areas. Furthermore, 'if out-migration reflects the locational preference of the population, it seems reasonable to suppose that it is of general benefit, leading to an improved quality of life for those who move from the city, and potentially for those who remain'.

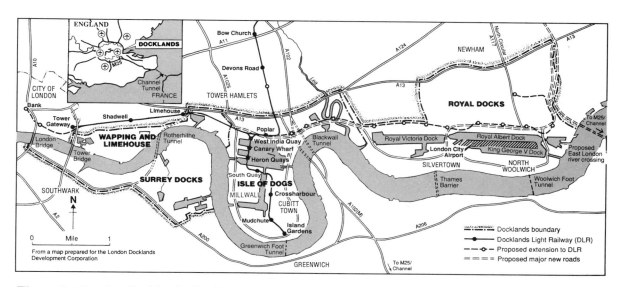

Figure 2.1 *London Docklands: Docklands covers 22 square km from London Bridge to Beckton. There are four main areas: Wapping and Limehouse, Surrey Docks, Isle of Dogs, and Royal Docks*

1 ISSUES AND CONTROVERSIES

Docklands is the largest urban renewal scheme in Europe. It extends eastwards from London Bridge to the Royal Docks at Beckton (Figure 2.1). A visit to this part of London's East End would reveal an area undergoing massive and rapid regeneration. However, will this *flagship programme* of the Government's inner city policy create a new and proud addition to London's landscape or, as some suggest, an environmental and social disaster which will ruin not only this part of the capital but much of the East End's hinterland?

Figure 2.2 *London Docklands: the exceptional place?* ▶

2   INNER CITY PROBLEMS IN DOCKLANDS

East London grew in the last century to serve the expanding dock system of the capital city. The area became 'a warehouse and a powerhouse' for Britain's growing international trade. In 1909 the Port of London Authority (PLA) was set up to manage the dock system of London from Tower Bridge to Tilbury. From 1967 onwards, faced with mounting difficulties, the PLA pursued a policy of shedding labour, closing the older upstream docks and concentrating investment on container handling in Tilbury. The last of London's up-

stream docks closed in 1981 (Table 2.1). There is no single reason for the closure of London's docks, rather a combination of forces from *global to local*. These include:

changing patterns and slump in seaborne trade;
changes in transport technology, such as the development of larger purpose-built vessels and containerisation, which the upstream docks were ill-equipped to handle;

Table 2.1 Docklands Factfile

Date	Dock closures	Main planning events	GLC changes
Phase I (1946–1967) Redevelopment			
1946		Comprehensive Development Areas eg Poplar	
1965			GLC set up, Conservative controlled
1967	East India Docks		
Phase II (1967–1977) Urban Experiments			
1968	London Docks	Urban Programme	
1969	St Katherine's Dock	Greater London Development Plan CDP Canning Town	Conservative retain GLC
1970	Surrey Docks	Competition for development. St Katherine's Dock	
1971		Hay's Wharf rezoned. Tower to Tilbury Conference	
1973		Travers Morgan Report	Labour GLC
1974		Docklands Joint Committee	
1976		London Docklands Strategic Plan	
Phase III (1977–1979) Enhanced Urban Programme			
1978		Docklands Partnership Area Docklands Development Organisation	Conservative GLC
Phase IV (1979 to date) Inner City Enterprise			
1980	West India Docks Millwall Docks South West India Dock		
1981	Royal Docks	London Docklands Development Corporation	Labour GLC
1982		Isle of Dogs Enterprise Zone	
1987		Docklands Light Railway London City Airport	GLC abolished
1988		Canary Wharf Development	

competition from other ports;

relatively high labour costs, labour disputes and management weaknesses;

low investment levels;

poor local infrastructure, such as inadequate cross-city transport links.

The ships left an area of 22 square km, with 88 km of waterfront and about 40 000 people. Closure brought inner city social and economic problems in its wake.

'The area displays more acutely and extensively than any other area in England the physical decline of the inner city and the need for urban regeneration'. (Secretary of State for the Environment, 1981)

2.1 Decline

i) Industrial decline

Despite its location in a generally prosperous region Docklands has experienced the full force of industrial restructuring, which has led to massive disinvestment and job loss. Apart from the docks the East End was traditionally dominated by small-scale manufacturers catering for the vast consumer market of London (clothing and furniture) and for the needs of the city (printing). By the end of the 1960s a threefold process of change was underway:

first, the *decline of the larger firms* in the manufacturing and transport sectors, such as the docks and the dock-related industries;

secondly, the *loss of jobs in small, labour intensive firms*, a consequence of changes in production, especially from foreign competitors who took full advantage of new technology, and changes in locational opportunities, with the growth of cheaper and more appealing sites on new, peripheral industrial estates;

thirdly, *the growth of services*, which signalled the encroachment of offices from the City eastwards.

The *rate* of these changes was not the same. Between 1961 and 1971 the Dockland Boroughs lost 106 000 jobs in manufacturing and transport, a loss of 3 per cent per annum. Over the next five years the rate speeded up to 4 per cent and by 1979 in Tower Hamlets the rate of decline was double that. The scale of decline in manufacturing jobs in Docklands has been particularly severe compared to London as a whole. Twenty five

per cent of all the jobs that left London between 1961 and 1971 were from the East End, although Docklands had only a quarter of manufacturing jobs in the first place. Gains in services were not able to compensate for this decline. In Tower Hamlets, in the same decade, losses in manufacturing amounted to 20 000 jobs, whilst increases in finance, insurance and banking totalled 4000. Table 3.2 (Part III) reveals the changing structure of employment in London's East End.

ii) Population decline

Like many other inner city areas London's docks have seen major changes in its population in recent years. The population declined as the young, better educated and better-off moved away (Table 2.2). In some cases the replacement population came from overseas. Docklands has traditionally been an area of new immigrant settlement. Throughout history it has hosted many different ethnic groups, including French, Chinese, Jews, Irish and most recently populations from the New Commonwealth and Pakistan. Often these groups concentrated into particular districts and streets. During the 1970s Brick Lane was settled by Bengalis. However, like many of the new arrivals in the past the East End's new Black population was vulnerable to the social and economic problems of the inner city.

Table 2.2 Population decline in Docklands

Borough	Population change (%)	
	1971–1976	1976–1981
Tower Hamlets	−22.0	− 4.0
Southwark	−25.3	−12.5
Newham	−21.6	− 7.8

Source: Census of Population.

2.2 Deprivation

By the mid 1970s Docklands was showing all the signs of the *multiple deprivation* typical of inner city areas.

Unemployment and poverty. The local communities of London's docks have long been used to social problems, most notably the poverty associated with casual labour. Dockers would queue for hours each morning for often dangerous and unpleasant work, without any guarantee of employment. Only as recently as 1967 was work de-casualised.

In recent years the cause of family poverty has been unemployment. In 1981 the male unemployment rate in Docklands was at 24 per cent, about twice the national average.

Poor housing. In 1981 more than 80 per cent of households lived in property rented from the local authority, mainly in large, poorly serviced inter-war council estates. One-third of these council houses were classed as *totally unsatisfactory* for human habitation. The remaining population were mostly found in tightly packed rows of small nineteenth century housing which had been hastily built and were now in need of drastic repair.

Derelict land. Derelict land was widespread. Close by, in the City of London, was some of the world's most expensive land, whereas property in Docklands was worthless. As docks and buildings became redundant they were often taken up by haulage and scrap merchants who controlled large sites at low rents and employed few people (Figure 2.3). In 1981 there was over 6 square km of vacant and derelict sites in Docklands.

Inadequate roads and public transport. Traditionally London's docks have only been accessible by rail or the River Thames. The docks were deliberately designed to be inaccessible to keep stores and bonded warehouses secure from thieves and pilfering. Cross-route links were not developed. Instead a maze of small roads sprang-up to feed particular wharves, which themselves were surrounded by high perimeter walls. In consequence, the outmoded infrastructure (roads, railways, drainage, telephones, electricity and so on) created a physically isolated and unattractive environment for new investment.

Run down community facilities. Docklands was increasingly served by a degenerating and ageing set of hospitals, schools, and welfare facilities. Declining catchments, administrative reorganisation and inadequate investment meant that in many cases closure was an inevitable outcome.

Multiple deprivation. According to a Department of Environment survey in 1981 Hackney, in Lon-

Figure 2.3 *Derelict site, Docklands: in 1981 there was over 6 square km of vacant and derelict sites in Docklands*

don's East End, was the most deprived area in England. Within its boundaries were:

> the highest proportion of single-parent families;
> the second highest level of overcrowding;
> the third highest concentration of New Commonwealth and Pakistani immigrants.

Within Greater London, Hackney had:

> the lowest average income;
> the highest proportion of decaying houses;
> the second highest level of violent theft in the Metropolitan Police Area.

2.3 Despair

Tension and social conflict have flared in Docklands. Throughout the 1970s bitter and often violent campaigns were staged against dock closures. In a famous study of Bethnal Green, Willmott and Young brought attention to the *declining community spirit* and *the weakening of neighbourly bonds*. An unfortunate development has been the growth of racism and racial attacks. The newly arrived New Commonwealth and Pakistani populations of the East End became the scapegoats for the inner city's problems. From 1971 onwards Brick Lane became the scene of organised racist violence against the resident Bengalis. In response Bengali self-defence groups were formed and a powder keg of social unrest was set alight.

3 THE REGENERATION OF DOCKLANDS: GOVERNMENT POLICY, 1946 TO DATE

The industrial decline and social change experienced by Docklands placed London's East End firmly on the political agenda in the debate about inner city policy. Should the existing structure of Docklands be improved, primarily to meet the needs of existing residents? Or should Docklands be transformed into a new commercial, financial and tourist centre, an extension of the West End, serving a different class of person? The answers to these questions have differed over the years. The replanning of Docklands has been influenced by three major factors:

first, changing explanations and solutions towards inner city problems in general;

secondly, the relative strengths of different local interest groups;

thirdly, shifts in local government organisation.

Table 2.1 summarises some of the principal planning events in Docklands.

3.1 Phase I (1946–1967)

Docklands suffered widespread bomb damage in the blitz of London. Like other large cities the earliest planning response was that of *Comprehensive Development*. One of the largest housing redevelopment sites in Britain was designated in Poplar in 1951. At the same time many Eastenders were encouraged to move to new and vast public housing estates under construction on the edge of London. From the mid 1960s policy solutions of this kind came under attack. It was argued that instead of relieving the problems of London's East End *regional policy was encouraging industrial decline and disinvestment.* Amongst the most vocal critics were local community and trades union organisations. As job losses proceeded, there was a growing lobby trying to press the government and local authorities to reverse their planning decisions.

3.2 Phase II (1967–1977)

In the late 1960s the *Greater London Development Plan* highlighted the serious nature of the industrial decline of London's East End. Presented with shrinking finances, as industrial ratepayers left the inner city, the local authorities soon added their voice to the critics of decentralisation. By the early 1970s local politicians were vigorously calling for *reinvestment* in Docklands.

These changes of opinion coincided with a *period of policy experimentation*. There was little agreement as to what sort of investment should take place in the area and where this should be. Sporadic area-based strategies resulted. The *Urban Programme* enabled voluntary and community groups to take advantage of special grants; *General Improvement Areas* and *Housing Action Areas* were designated; and a *Community Development Project* was launched in Canning Town.

Of special significance were the schemes to develop St Katherine's Dock and Hay's Wharf. In 1970 the Conservative-run Greater London Council (GLC) bought St Katherine's Dock from the PLA. A competition was held for its redevelopment. They chose a proposal which included a hotel, trade centre, marina and public and private housing (Figure 2.4). On the other side of the river Hay's Wharf was *re-zoned as West End*, opening the way for office development. Both schemes established a pattern of private development which has become a principal feature of modern Docklands.

Figure 2.4 St Katherine's Dock, converted to new uses after its closure in 1969

Both projects sparked off local opposition as they were seen to be inappropriate to the needs of Eastenders, providing opportunities only for outsiders and the better-off. In response, the *Tower to Tilbury Conference* held in 1971 proposed that the upper Thames should concentrate on housing and jobs. Planning consultants Travers Morgan were brought in to draw up proposals for East London. They suggested five options, the most favoured called for a fundamental change of use for the area. In effect they wanted to create a *City New Town* by bringing the West End to the East End.

In 1973 the incoming Labour controlled GLC, aware of local opposition, shelved the *Travers Morgan Report* and set up the *Docklands Joint Committee*. This group comprised of representatives from the GLC and the five Dockland Boroughs. After extensive public consultation,

London Docklands: A Strategic Plan was released in 1976. The report emphasised the need to stem existing job losses, to bring new jobs that would match local skills, to use the vacant land to tackle the acute housing problems and to make the area attractive. It was acclaimed as a *comprehensive needs-led solution* and a significant step forward for local people.

3.3 Phase III (1977–1979)

Under the 1978 *Inner Urban Areas Act* Docklands became a *Partnership Area* and so eligible for additional funding. However, implementation of the Strategic Plan was beset with difficulties. Public spending cuts, which starved the scheme of funding, and difficulties in acquiring land for development from public agencies like the PLA, gas, electricity and water boards, meant that the Plan's original goals were blocked. Also, the newly appointed Conservative GLC were not committed to the Plans' objectives, because there was too much bureaucratic control over, and insufficient stimulation of, private investment. In 1978 they set up the *Docklands Development Organisation*, a promotional agency with no local accountability. The drift away from needs-led solutions gathered momentum.

3.4 Phase IV (1979 to date)

In 1981 the Government set up the *London Docklands Development Corporation* (LDDC), an *Urban Development Corporation*, and gave it the task of bringing about the economic, social and physical renewal of the Docklands (Figure 2.1). This was the beginning of a new era for London's East End. The LDDC is a *non-elected, single purpose, planning and approval agency*. It has total control of the development process in the Boroughs of Tower Hamlets, Newham and Southwark, though it is not the housing, education or health authority. The LDDC does not undertake redevelopment directly. It uses its substantial funds (from Government grants and land sales) to acquire and prepare land and then release it to developers, largely in the private sector. In other words, its main approach has been *to use public funds as a lever for private enterprise*. This approach, which encourages the private investor to set up what they consider to be the most profitable land-uses, is known as *market-led* or *demand-led planning*. A major consequence of such

Figure 2.5 Flagship projects: Canary Wharf. How 'New Manhattan' will look

an approach is that unlike earlier strategies there is no overall plan for Docklands development.

Development has proceeded apace in Docklands, notably in the private and highly expensive residential sector and most recently, in the office and commercial sector. The LDDC claims that over the period 1981–1988 about £440 million of public investment has encouraged £4400 million of private investment, a leverage ratio of 1:10 (see Finance section below). Within a decade the whole face of London's East End has been completely transformed, but questions have been raised about whose interests the regeneration of Docklands is serving. Local community groups point out that the people of Docklands have gained little from the Corporation's demand-led strategies.

i) Flagship projects

Docklands is portrayed as a model approach of what can be done to regenerate the inner city. The most ambitious developments are at Canary Wharf and the Royal Docks.

Canary Wharf: the New Manhattan. By far the largest changes have taken place on the Isle of Dogs, part of which was designated an *Enterprise Zone* in 1982. The private developers are Canadian based Olympia and York (OY). They have invested more than £3700 million into a scheme which involves the construction of 24, super-scale, office and commercial buildings with up to 12 million square feet of floor space, where 50 000 people will be employed (Figure 2.5). At the centre of Canary Wharf will be Britain's tallest tower block, an 800 foot skyscraper. The scale and extent of this programme, known as New Manhattan (or *Wall Street on Water*), makes it larger than any other in Europe. Along with three smaller developments at nearby North Quay, South Quay and Heron Quays, Canary Wharf will bring in a daytime population equivalent to the size of Cambridge. When Canary Wharf is finished in 1993 there will be 120 000 workers travelling everyday into the Isle of Dogs.

Royal Docks. The LDDC believes that the Canary Wharf scheme has created a 'bow wave of momentum', providing massive private interest in the proposed development of the Royal Docks. Redevelopment of the Royal Docks is the Corporation's last major project. Hailed as *London's Water City*, the £2 billion proposal includes a vast regional shopping centre, 7000 new houses, a business/high-tech park and an exhibition centre.

ii) Catalyst schemes

Changing the reputation and image of Docklands has been a key feature of LDDC policy. This has involved high profile marketing and the use of catalyst projects, major developments which attract further private investment and help change perceptions of the area. Transport improvements are seen as highly important in this sense.

Docklands Light Railway. In order to improve access to the area the Docklands Light Railway was opened in 1987 at a cost of £77 million (Figure 2.6). Unmanned and programmed to stop by computer, it connects the Isle of Dogs to the main rail and underground network and the City of London. Plans exist to extend the route eastwards to the Royal Docks by the early 1990s.

London City Airport. A further boost was given to Docklands with the opening of the London City Airport (formerly *STOLport* – Short Take Off and Landing) in 1987. This has been developed on the former quay between the Royal Docks at Beckton. Brymon Airways and London City Airways offer scheduled flights to domestic and European destinations within a 600 km radius of London.

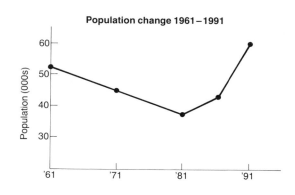

Figure 2.6 *Catalyst projects, new transport schemes: Docklands Light Railway*

iii) Economic, social and environmental change

Figure 2.7 shows some of the principal changes taking place in Docklands. Not all of these changes have been welcomed. In the following discussion we shall examine the advantages and disadvantages of the development process.

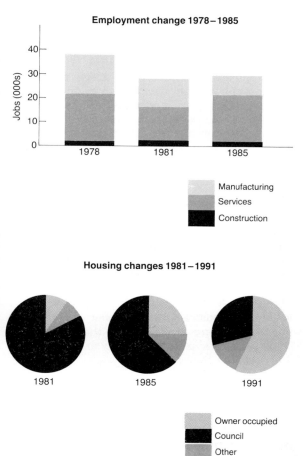

Figure 2.7 *Social and economic change in Docklands*

Advantages	Disadvantages

Employment

Slogans like 'why move to the middle of nowhere when you can move to the middle of London' have been used to draw attention to the changing image of Docklands.

Between 1981 and 1987, 8000 new jobs have been added to the Docklands economy, a rise of 34 per cent. By 1991 a further 25 000 jobs are expected. The jobs are mainly in three sectors: finance and business services; distribution, hotels and catering; printing and publishing (Figures 2.7 and 2.8)

Of the 8000 new jobs, 5000 have been transferred from elsewhere (firms bringing their workforce with them). Only 3000 jobs are new in the sense that they did not exist before (Figure 2.8). The new jobs do not match local skills and so are filled by people from outside the area. Only 13 per cent of the workforce of incoming firms is local. Job losses in the traditional sectors continue. Since 1981 13 000 local jobs have been lost.

In 1988 52 000 residents were unemployed in Docklands, a number higher than 1981.

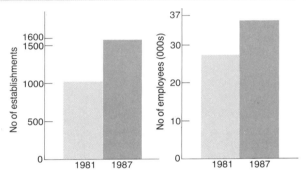
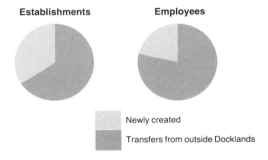

Figure 2.8 *Employment change in Docklands, 1981–87*

Housing

Huge house building programmes have been promoted by the LDDC, principally at Beckton, Wapping, Surrey Docks and the Isle of Dogs. In 1988 nearly 9000 new houses had been built, and work had started on over 15 000. A further 8000 are planned. River frontage sites are popular with purchasers new to the area.

Over 80 per cent of housing is for private sale, mostly fetching prices well above what local people can afford. In Newham only 4 per cent of new private sales are to locals. Rising prices compound the problem. At the top end of the market prices of £1 million for penthouses have been recorded. Over two-thirds of the land given over to the LDDC in 1981 had been earmarked for local authority housing.

The high-tech, Lego-land style of architecture favoured in Docklands is criticised for being out of keeping with the locality.

Figure 2.9 *New housing developments: Clippers Quay, Isle of Dogs*

Advantages	Disadvantages
Population	
Since 1981 the population of Docklands has been growing and is expected to rise throughout the 1990s (Figure 2.7).	Both the growth of the business sector and private house sales have attracted a new population into Docklands. The movement in of a young, upwardly mobile and professional population (*Yuppies*) has transformed the social geography of Docklands. Friction between the old and new populations is mounting.
Transport	
New catalyst schemes have opened up Docklands: for example, the Docklands Light Railway and STOLport (Figure 2.10). One of the LDDC's earliest projects was a network of red brick roads in the EZ, linking the Isle of Dogs to the City. The *Docklands Highway* is the newest scheme: this includes the Limehouse link, a cut and cover road. In 1988 a high-speed river-bus service, Thames Line, was introduced to serve commuter and tourist traffic.	The capacities of existing and planned schemes are unlikely to meet the passenger demands created by new developments like Canary Wharf. Costs are escalating. The cost of the Limehouse Link has risen from £85 million to £140 million before work has started. Many council houses are being demolished in the wake of the new road schemes. Too much emphasis is being given to road developments at the expense of public transport.

Figure 2.10 *London City Airport (formerly STOLport)*

Local democracy and community programmes	
Since 1987 the LDDC has increased its commitment to social and community aspects of regeneration. Two positive arrangements have been reached. The first is the *Memorandum of Agreement* signed between Newham Council and the LDDC in 1987	From the outset the LDDC faced criticism from local boroughs, community groups and organiations like the Docklands Community Council (DCC), for its lack of democracy, local consultation and communication. The LDDC is accused of showing

Advantages	Disadvantages
over the Royal Docks development. This promises 1500 'social houses', 25 per cent of all new jobs to be made available to local residents and provision of social facilities.	inadequate concern for the needs of locals. In their document *Local Democracy Works: a Partnership in London Docklands*, the DCC criticise the LDDC's ability to impose changes without public meetings (Figure 2.11).
Secondly, Tower Hamlets Council has struck two deals: one with the LDDC over the rehousing of council tenants affected by the Limehouse Link; the other with OY, guaranteeing 2000 Canary Wharf jobs and £2.5 million towards training for local people.	

Finance

The LDDC and the EZ have attracted approximately £440 million in grants since 1981. This has allowed the LDDC to offer an array of incentives and subsidies to developers, including cheap land, prepared sites and provision of infrastructure.

At the same time as record public subsidies are going to the LDDC, local councils have had their resources cut back by rate capping and capital spending controls.

Recent doubts have been cast on the LDDC's ability to meet its widening commitments. An independent report suggests a public to private leverage ratio of 1:4 (not 1:10, see above), when sums from all public bodies are totalled.

In 1988 the Corporation asked the Government for an urgent increase in funding to stave off a possible deficit of £404 million over the next four years.

With the same amount of money as the LDDC, local councils claim they could have done a much better job.

Big money is moving in

Massive luxury riverside development – public housing cut to nil.

Large enterprises moving in (bringing their own workforce) get incentives while small local firms are squeezed out – few new jobs will be created.

Don't let it push out local people

Figure 2.11 *Local action groups are unhappy about the ways in which Docklands is being developed. The campaign slogan reproduced here was used on a community poster in Wapping entitled 'The Changing Picture of Docklands'*

Planning

Docklands is Europe's biggest building site. There is no question that the pace and scale of development in the last decade has been colossal. There is a real sense that things are happening.

What Docklands needs is a plan. The idea that the market is the answer to everything needs to be closely examined. Olympia and York have a master plan for Canary Wharf, but there is no overall plan for the whole of Docklands. At the moment there is little attempt to regulate the flow of development. Eventually the building work will stop, but at what cost to the environment and to the Eastenders?

4   REGENERATION OF DOCKLANDS: SUCCESS OR FAILURE?

The regeneration of Docklands is seen by the present Government as a model of its approach to the inner city. For example, policy decisions are based firmly *on central rather than local decisions, on private rather than public sector investment*, and *on economic rather than social concerns*. Indeed Docklands has been described as 'a test-bed of government policy; much of what we experience today, others experience tomorrow' (DCC, 1988). However, in pure development terms Docklands appears a *one-off*. The demand for commercial and private residential development on such a scale is unlikely to be matched, for instance by the Merseyside Development Corporation. The position of Docklands, next to the City of London and in the prosperous South-east, is an unrepeatable advantage.

Whether local people will reap the benefits of the enormous changes taking place in Docklands remains to be seen. At present there is little sign of the hoped for *trickle-down effect*, whereby somewhere along the line everyone will benefit either directly or indirectly from market-led development. Instead 'two Dockland environments now exist side by side, one luxurious enclaves for wealthy incomers, another underfunded estates, parks and community facilities where residents live' (DCC, 1988) (Figure 2.12).

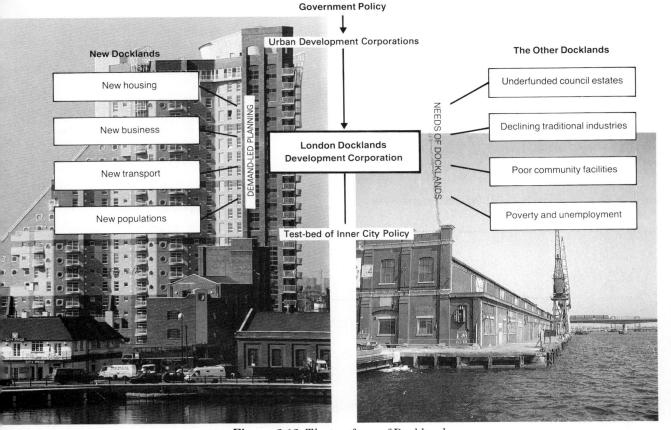

Figure 2.12 *The two faces of Docklands*

In this Part different types of exercises are provided: some involve data response, others depend upon fieldwork and data collection. All of the work can be carried out individually, although suggestions for group activities are made. This section has been written to complement Parts I and II, and a reading of these earlier Parts is recommended before any of the exercises are attempted.

1 DECLINE

1.1 Population decline

Hypothesis: The inner city is an area of disproportional population decline

Study Table 3.1

i) Which Boroughs show the largest population loss? Which Boroughs show population gain?

ii) a) Group the percentage data (column 3) into classes of similar values. Figure 3.2 shows how classes can be distinguished by plotting a dispersion diagram. Try to get no more than 5 classes to represent either the negative or positive values.

b) Choose a shading category for each class.

c) Copy Figure 3.1. Draw a choropleth map to show patterns of population change in London.

Table 3.1 Population change in Greater London, 1951–81

London Borough	Population (thousands)		% Change
	1951	1981	1951–1981
Barking	189	150	−21
Barnet	320	293	− 8
Bexley	205	215	+ 5
Brent	311	253	−19
Bromley	268	297	+11
Camden	258	172	−33
Croydon	310	318	+ 3
Ealing	311	280	−10
Enfield	288	259	−10
Greenwich	236	212	−10
Hackney	265	180	−32
Hammersmith	241	148	−39
Haringey	277	204	−26
Harrow	219	197	−10
Havering	192	241	+26
Hillingdon	210	230	+10
Hounslow	211	201	− 5
Islington	271	161	−41
Kensington & Chelsea	219	139	−37

London Borough	Population (thousands)		% Change
	1951	1981	1951–1981
Kingston upon Thames	147	133	−10
Lambeth	347	246	−29
Lewisham	303	231	−24
Merton	200	166	−17
Newham	294	209	−29
Redbridge	257	226	−12
Richmond upon Thames	188	160	−15
Southwark	338	212	−37
Sutton	176	169	− 4
Tower Hamlets	231	143	−38
Waltham Forest	275	216	−21
Wandsworth	331	255	−23
Westminster	300	191	−36
GREATER LONDON	8 197	6 713	−18
Rest of England & Wales	35 561	42 315	+19

Source: Census of Population.

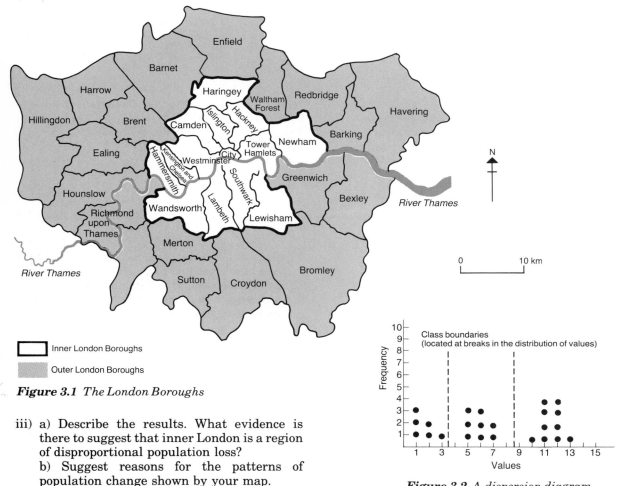

Figure 3.1 *The London Boroughs*

iii) a) Describe the results. What evidence is there to suggest that inner London is a region of disproportional population loss?
   b) Suggest reasons for the patterns of population change shown by your map.

1.2 Industrial decline

Hypothesis: The inner city is an area of declining job opportunity and changing employment structure

Study Table 3.2

i) Draw proportional pie-charts to show the size and structure of employment in Tower Hamlets in 1961 and 1974.
ii) Describe the changes in employment shown by the pie-charts.
iii) Name and explain *three* social and/or economic problems which inner cities are likely to experience as a result of such employment change.

Figure 3.2 *A dispersion diagram*

Table 3.2 Employment size and structure in the London Borough of Tower Hamlets

	Number of jobs	
	1961	1974
Manufacturing	57 000	29 500
Construction	7 300	5 000
Services	14 000	36 800
Distribution	21 450	11 000
Transport	33 250	19 700
Total	133 000	102 000

Source: London Borough of Tower Hamlets.

2   DEPRIVATION

2.1 The nature of inner city deprivation

Hypothesis: The inner city is a difficult region to define, its form and extent varies from place to place

Hypothesis: The inner city is a region of multiple social and economic deprivation

Study Table 3.3

i) Select four indicators from Table 3.3 to represent different aspects of social and/or economic deprivation. Justify your choice.

ii) Rank each indicator from high to low relative deprivation: a rank value of 1 represents high deprivation and a rank value of 18 represents low deprivation. Take care to rank the values in an appropriate manner.

iii) a) For each ward add together the four rank values to provide a total rank score.
b) Which wards have the highest and lowest total rank scores? Where are these wards located?
c) Which wards have high deprivation scores on more than one indicator?

iv) a) Group the total rank scores into no more than five classes (see 1.1 ii above).
b) Copy Figure 3.3. Draw a choropleth map to show the distribution of these classes.

v) a) Describe the patterns shown by your map. Is there a region of multiple social and economic deprivation which can be distinguished as the inner city? If so, what is its form and extent?
b) How does this region differ in location and form to *the zone of transition* in Burgess' model of the city?
c) Suggest reasons for any difference.
d) Compare your map with Figure 1.1, which shows the designated *inner city* (Priority Area) in Coventry.

Table 3.3 Social and economic indicators for wards in Coventry, 1981

Wards	% population change 1971–81	% persons in households with NCWP[1] born head	% unskilled workers (SEG 11)	% unemployed	% households without a car	% households with >1 person per room[2]	% households without indoor wc	% households owner occupied	% households rented from private landlord
Bablake	− 7.8	1.8	5.5	9.4	31	2.3	0.2	81	6
Binley and Willenhall	+ 4.9	3.6	6.6	15.8	42	5.8	1.3	49	5
Cheylesmore	− 6.9	5.7	6.0	11.7	35	4.0	1.2	76	10
Earlsdon	−10.7	3.4	2.2	8.9	29	1.3	1.5	82	12
Foleshill	−17.1	50.2	15.6	23.0	62	15.8	7.5	59	27
Henley	− 7.1	3.9	8.4	18.9	47	8.2	0.5	46	6
Holbrooks	− 3.1	11.0	7.3	13.9	40	8.0	3.7	71	12
Longford	− 6.4	10.2	8.5	15.1	48	6.7	6.0	61	10
Lower Stoke	−11.0	7.5	8.9	14.2	49	5.7	2.2	68	9
Radford	−11.4	12.3	11.6	17.5	54	7.8	4.0	53	13
St Michaels	−15.4	29.9	12.5	22.0	68	10.0	6.8	45	25
Sherbourne	− 9.4	3.4	5.6	12.4	40	3.6	1.4	77	12
Upper Stoke	−10.4	12.0	8.1	15.4	50	6.4	3.5	69	9
Wainbody	+ 4.8	2.9	1.8	6.2	16	1.2	0.4	92	4
Westwood	−11.8	1.5	7.2	16.7	47	4.7	1.4	51	7
Whoberley	−10.5	3.7	4.3	12.3	40	3.4	2.6	81	11
Woodlands	− 1.5	1.2	4.2	11.6	32	3.0	0.4	65	9
Wyken	+ 5.5	6.3	5.1	9.4	34	4.6	0.8	74	6

1 New Commonwealth and Pakistan. 2 A measure of overcrowding.   Source: Census of Population.

Figure 3.3 *Wards of Coventry*

2.2 Aspects of deprivation

Hypothesis: The inner city is a region of comparative poverty; the people who live in the inner city are poorer than those who live in the suburbs

Hypothesis: The inner city is a region of poor housing conditions and standards

Hypothesis: The population of the inner city comprises higher proportions of children and the elderly than elsewhere in the city

The information shown in Table 3.3 was found in the Small Area Statistics of the 1981 Census of Population. The Census provides a lot of statistical information about urban areas, mostly in tabulated form. Many projects can be designed around the data found in the Census. Censuses are avail-

able for public view at most large libraries, County Archive Offices and City Planning Departments.

i) For your nearest large city select ward data (for London use Boroughs) from the Census to test each hypothesis.

Table 3.3 shows some indicators which will be of value to your study. Other indicators include:

% households renting from the Council
% households lacking bath/bath and inside wc
% households sharing bath
% households with 2 or more cars
% households with more than 6 people (large households)
% of children under 16 in households with only one parent (lone parent households)

% of persons aged 0–15
% of persons of pensionable age

ii) a) In each case draw a choropleth map to highlight any spatial variations in the data.
b) Describe the patterns shown on your finished maps.
c) How useful has your choice of information been in testing each of the hypotheses?
iii) a) Describe *two* other ways of testing each of the hypotheses.
b) Using these two techniques, show whether each hypothesis is true, partly correct or incorrect.
iv) a) For larger cities only*, classify the wards *by location* into inner and outer city wards.
b) This classification can be used as a framework for *two-sample, chi-squared analysis* (X^2). An example of a contingency table is shown in Table 3.4.
c) Do the X^2 values confirm or support the hypotheses?

Table 3.4 Contingency table for two-sample chi-squared analysis

| | Housing deprivation scores (any suitable indicator can be substituted) | | |
	High	Medium	Low
Inner city			
Outer city			

*Expected values in 20% of the cells must not be <5

$$X^2 = \sum \frac{(O-E)^2}{E}$$

where X^2 = chi-squared
$\sum$ = sum of
O = observed frequency
E = expected frequency

Test the calculated values of X^2 for significance: calculate degrees of freedom (df)
df = (r−1) (k−1)

where r = number of rows
k = number of columns

Refer to significance tables of X^2. Use the calculated value of X^2 and df to check the level of significance (p) of your results.

2.3 Environmental and housing quality

Hypothesis: The inner city is a region of poor housing quality and environmental decay. Better housing and more pleasant environments are found in the suburbs.

i) In order to test this hypothesis you will need to undertake a field survey.
Table 3.5 suggests a classification for assessing environmental quality. Penalty points are awarded to each location according to relative levels of deficiency. The maximum number of penalty points is 60. The total is divided between different elements of the environment. This allows the variable nature of different locations to be assessed.
ii) Two different strategies of data collection are suggested: the choice will depend on whether you are working alone or as part of a group and the size of the urban area.
Strategy 1: Superimpose a grid of convenient size upon the urban area. You could use the OS 1 km grid as a starting framework. Each grid square can then be subdivided into quarters or even more finely. At the mid-point of each cell (or as near as possible) measure the quality of the urban environment using the proposed classification.
Strategy 2: If the urban area is large, concentrate on selected areas from the outset. Choose a part of the inner city and contrast with a suburban location. In this case measurements can be recorded by street or a fine grid can be superimposed on both study areas, like above.
iii) The analysis should focus on whether variation in environmental quality exists and of what kind between the inner city and the suburbs. Think of different ways of displaying the data. For example, values could be recorded at the mid-points of each cell and lines of equal environmental quality interpolated to form an *isopleth map*. The data could be grouped into classes and the cells shaded to create a *choropleth map. Divided bar graphs* can be drawn for different locations, representative of both the total penalty score and its components.

Table 3.5 A classification for assessing environmental quality

Environmental quality	Penalty points
Housing conditions	
i) Acute physical decay: many signs of deterioration and disrepair (eg missing slates, broken windows, crumbling brickwork, boarded-up spaces)	4–5
ii) Some signs of deterioration and disrepair	1–3
iii) No signs of deterioration and disrepair, housing in good condition	0
Appearance of gardens/yards	
i) Most gardens/yards unkempt and poorly fenced/walled	4–5
ii) Some gardens/yards unkempt and poorly fenced/walled	1–3
iii) Most gardens tidy, well-maintained and neatly fenced and walled	0
Garaging/parking provision	
i) 0–24% provision of garages: considerable on-street parking	5
ii) 25–49% provision of garages	4
iii) 50–74% provision of garages	3
iv) 75–95% provision of garages	2
v) Full provision of garages: very limited on-street parking	0
Landscaping	
i) Total, or almost total, lack of trees/grassed spaces	4–5
ii) Poor quality trees: poorly placed and unkempt grassed spaces	1–3
iii) Mature, good quality trees: well placed and well kempt grassed spaces	0
Townscape	
i) Excessive drabness of built environment	4–5
ii) Some drabness of built environment	1–3
iii) Attractive built environment	0
Dereliction	
i) Evidence of abandoned land: transport (eg railway, waterside), industrial (eg gravel, manufacturing), cultivated (eg allotments), residential	4–5
ii) Evidence of transitional land (development land): urban renewal, industrial development, road building	1–3
iii) Well established land uses	0
Traffic	
i) Substantial intrusion by through traffic	4–5
ii) Some intrusion by through traffic	1–3
iii) No intrusion by through traffic	0
Noise	
i) Above acceptable residential standard: normal speech difficult at times	4–5
ii) Some intrusion by noise: limited speech interference	1–3
iii) Acceptable residential standards	0
Air pollution	
i) Heavy	4–5
ii) Light	1–3
iii) Negligible	0
Litter	
i) Excessive litter	4–5
ii) Some litter	1–3
iii) Limited amounts of litter	0
Graffiti	
i) Much evidence	4–5
ii) Some evidence	1–3
iii) No graffiti	0
Provision of neighbourhood amenities (street lighting, telephone kiosks, post boxes and bus stops)	
i) Total, or almost total, lack of neighbourhood amenities within five minutes walking distance	4–5
ii) Some amenities absent within five minutes walking distance	1–3
iii) Full provision of neighbourhood amenities within five minutes walking distance	0

3   DESPAIR

3.1 Quality of Life

Hypothesis: The quality of life in the inner city is lower than in the suburbs: a feeling of despair is evident in the inner city

In order to uncover the attitudes and opinions of people who live in the inner city compared to those who live in the suburbs you will need to carry out a questionnaire survey.

i) People's sense of satisfaction with their lives and their overall sense of well-being can be assessed by looking at seven important aspects of life (Table 3.6). Respondents are asked to allocate a score to each aspect of life within a set scale.

ii) Satisfactions with aspects of the local district can be measured in the same way (Table 3.7).

iii) Levels of participation and alienation can be assessed by asking some of the following questions:

Can you name your local councillor? (Yes/No)

Have you ever contacted (in person, by letter or telephone) a local councillor or public official regarding a local issue? (Yes/No)

Have you ever signed a petition regarding any problem in this area? (Yes/No)

Do you usually vote in local elections? (Yes/No) If not, why?

Do you feel that you belong to this local area? (0−no feeling of belonging to 4−strong sense of belonging)

When you think of this local area what problems come to mind?

Are you seriously thinking of moving out of this area? (Yes/No)

How long do you intend to stay in your present accommodation? (<1 year, 1−5 years, >5 years, indefinitely)

iv) People's sense of satisfaction may change over time (Table 3.8).

v) a) Compare the results of each area.

b) Calculate the mean scores for each aspect of life, each aspect of the local area, overall quality of life now, then and in the future.

c) Do people respond positively (values >2) or negatively (values <2) to these different aspects of their life and local area?

Table 3.6 Perceived quality of life

Aspect	Points
Job	0−4
House	0−4
Local area	0−4
Public transport	0−4
Primary education	0−4
Secondary education	0−4
Health services	0−4

Table 3.7 Satisfaction with aspects of the local area as a place to live

Aspect	Points
Services	0−4
Shops	0−4
Peace and quiet	0−4
Appearance and tidiness	0−4
Opportunities for entertainment	0−4
Levels of crime/vandalism	0−4
Personal safety/security	0−4
Reputation of the area	0−4
Friendliness of area	0−4
Neighbours	0−4
Opportunities for employment	0−4
Pollution	0−4

Table 3.8 Changing sense of satisfaction with the local area

Aspect	Points
Now	0−4
Five years ago	0−4
Five years ahead	0−4

Scale
0 Not relevant
1 Extremely dissatisfied
2 Dissatisfied
3 Satisfied
4 Extremely satisfied

4   POLICY

4.1 Living in Docklands

New residential property in Docklands is often expensive. The high cost of such property, together with its size, facilities and location make it appealing to particular types of purchaser.

i) Why do people *choose* to live in Docklands as opposed to suburban locations, and pay so much to do so?

ii) How are these purchasers likely to differ from the *established* Eastenders in terms of: age; marital status; income; occupation; likely length of future residence in the area; job mobility; life style; use of local services and facilities?

iii) a) What term(s) is usually used to describe the process of social change which comes about through the development of expensive property in formally run down places?
b) Explain the meaning of this term.

iv) Place yourself in the role of each of the following characters:

wine-bar owner;
estate agent;
unemployed, former dock worker, council house tenant;
merchant banker;
Bengali woman, clothing (*sweatshop*) worker, living in old rented accommodation.

a) What are their reasons for living in Docklands?
b) How much choice do you think each character has with regard to their place of residence?

v) Imagine what each of these characters would think of the following developments in Docklands:

Canary Wharf development;
STOLport and Docklands Light Railway;
growth of luxury housing;
low funding of council estates;
the role of the LDDC.

vi) Suggest what sort of changes each of these characters would most welcome in Docklands in the next decade. Use the following headings to organise your discussion: social

change; economic change; environmental change.

4.2 Planning Docklands

This exercise involves a simple role-playing game. You can either carry out this exercise by yourself or in small groups, with each group assuming a different role.

Scenario: The Royal Docks are set to become the focal point of Docklands development during the 1990s. Figure 3.4 shows current developments and vacant sites in the Royal Docks.

i) Your task is to prepare a development plan for the Royal Docks. Such plans are rarely straightforward to devise. Different groups and organisations will often have very different ideas as to what should go ahead. For this exercise five individuals, representing different interests, have been asked for their opinions (Table 3.9).

ii) a) For each role, select *three* land-uses which you consider to be most appropriate to go ahead in the Royal Docks. Possible land-uses include: luxury private housing; low-cost private housing; council housing; hotels; sports halls; exhibition halls; marinas; parks; car parks; offices; industry; small shops; large retail park etc.
b) Justify your choice of land-uses;
c) How do the land-uses chosen by each character differ?
d) For each role, on a copy of Figure 3.4, locate these land-uses.
e) Justify your choice of location.
f) Compare and contrast the land-use maps.

iii) If the exercise is being carried out by small groups you can develop the role play further.
a) Appoint a chairperson. This person should assume the role of being a member of the LDDC. What criteria for development do you think such a person would consider important?
b) Acting out your roles, take your plans to a meeting and work towards devising a strategy for the whole area.
c) Make a note of how decisions are reached. Was each *individual* satisfied with the final plan?

Figure 3.4 *The Royal Docks*

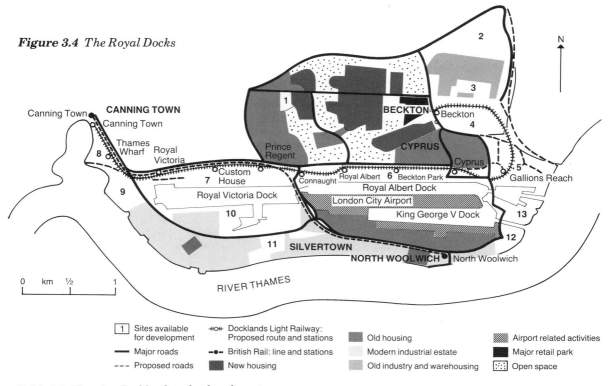

Table 3.9 Planning Docklands: role-play characters

D. Smith, Executive Property Developers
You represent a large property developer. The firm's main interest is to make as much profit as possible. The Royal Docks offers ideal sites for expensive property of different sorts. Your company has the money to develop the area. You are not concerned about the needs of the established residents, but you are aware of growing opposition to private enterprise.

P. Nicholls, Eastenders Action Group
You have lived in the East End all your life (50 years). You head an action group set up to represent the needs and aspirations of the local people. You want development that will benefit local Eastenders, not newcomers. However, you recognise that development is unlikely to go ahead without private capital.

D. Clark, Wapping Council
You are leader of the Labour controlled local authority. You agree that the Royal Docks require redevelopment, but appreciate that the local authority has insufficient finance to carry through the full process of change. You realise that new developments are likely to bolster the tax-base of the area and so bring in more money and opportunities. You want the sort of development that would be attractive to private developers and yet benefit the whole community.

R. Jones, Conservative MP for City
You are a Conservative Member of Parliament for the nearby City constituency. London Docklands is currently represented by a Labour MP. The Government would like to see the area become a safe Conservative seat. The local Conservative Party are pleased to recruit your support for their campaign committee. You have considerable influence with private developers and prepare plans which would radically change the social and economic character of the area.

A. Browne, Leisure Trust
You represent a group putting the case for the leisure industry. Royal Docks provides large waterside sites with nearby access to the centre of London. You would like to open up new leisure opportunities in the area, which would appeal to as wide a market as possible. However, you realise that the leisure developers want to make good profits in their investments.

SUGGESTED READING

There are many good books on the inner city, some are suggested below:

Archbishop of Canterbury's Commission (1985) Faith in the city. London, Church of England Commissioners

Clark, D. (1989) Urban decline. London, Routledge and Kegan Paul

Donnison, D., and Middleton, A. (1987) Regenerating the inner city: Glasgow's experience. London, Routledge and Kegan Paul

Hall, P. (1981) The inner city in context. London, Heinemann

Harrison, P. (1983) Inside the inner city: life under the cutting edge. London, Penguin Books

Lawless, P. (1989) Britain's inner cities: problems and policies. London, Harper and Row

Robson, B. (1988) Those inner cities. Oxford, Clarendon Press

INDEX